Digital Storage
Oscilloscopes

Digital Storage Oscilloscopes

Ian Hickman BSc (Hons), CEng, MIEE, MIEEE

Newnes
An imprint of Butterworth-Heinemann

Newnes
An imprint of Butterworth-Heinemann
Linacre House, Jordan Hill, Oxford OX2 8DP
A division of Reed Educational and Professional Publishing Ltd

A member of the Reed Elsevier plc group

OXFORD BOSTON JOHANNESBURG
MELBOURNE NEW DELHI SINGAPORE

First published 1997
© Ian Hickman 1997

All rights reserved. No part of this publication may be reproduced in any material form (including photocopying or storing in any medium by electronic means and whether or not transiently or incidentally to some other use of this publication) without the written permission of the copyright holder except in accordance with the provisions of the Copyright, Designs and Patents Act 1988 or under the terms of a licence issued by the Copyright Licensing Agency Ltd, 90 Tottenham Court Road, London, England W1P 9HE. Applications for the copyright holder's written permission to reproduce any part of this publication should be addressed to the publishers

British Library Cataloguing in Publication Data
ISBN 0 7506 2856 1

Library of Congress Cataloguing in Publication Data

Designed, typeset and produced by Co-publications <copubs@aol.com>

Printed and bound in Great Britain by Hartnolls Ltd. Bodmin, Cornwall

Contents

	Preface	vii
1	Introduction	1
2	Digital storage oscilloscope fundamentals	6
3	Digital storage oscilloscope modes of operation	13
4	Input circuitry	26
5	Digitising	33
6	Triggering	42
7	Signal processing	49
8	Displays and display technology	67
9	Digital sampling oscilloscopes	84
10	Accessories for use with digital storage oscilloscopes	103
11	Using digital storage oscilloscopes	120
	Glossary	139
	Appendix	144
	Index	148

Preface

Digital storage oscilloscopes first appeared around a quarter of a century ago. At that stage their performance was limited, not even approaching that of the analog (cathode ray tube) storage oscilloscopes of the day in many respects, such as bandwidth, in particular. Nevertheless, they offered certain facilities not available in the earlier analog storage oscilloscopes, principally pre-trigger storage, permitting the user to capture part of a waveform which preceded a trigger event. Since those early days, development has been continuous and intensive, with the result that digital storage oscilloscopes have long since completely supplanted analog storage oscilloscopes and have indeed reached a degree of sophistication and performance which enable them to rival the most advanced real time oscilloscopes.

This being so, the time seemed right for a book dealing solely with digital storage oscilloscopes, and this work is the result. It may be seen as a companion volume to *Oscilloscopes: How to Use Them, How They Work*, 4th Edition 1955, ISBN 0 7506 2282 2, by the same author, which deals with oscilloscopes and waveform recorders of all sorts. But being devoted exclusively to digital storage oscilloscopes, it is able to cover the ground in more detail than the book just mentioned, and also to cover more recent models of digital storage oscilloscope which had not yet been released when the earlier book appeared. For such is the bewildering speed of development in oscillography, as in electronics generally, that any book on the subject is bound to be partially out of date by the time it appears in print. Nevertheless, the basic principles of the digital storage oscilloscope remain the same, and the user will gain a great deal of knowledge about, and insight into the workings of, these versatile instruments from this book.

Ian Hickman
January 1997

Chapter 1

Introduction

Digital storage oscilloscopes have been around for many years. They first appeared in the early 1970s, and at least two companies claim to have been the originators. However, as these two companies are now part of the same group, the question is largely academic. What is not in question is the fact that, since those early days, the capabilities of digital storage oscilloscopes have been developed to a degree that could not then have been dreamed of, to the point indeed where some manufacturers claim that they can usefully replace traditional analog oscilloscopes entirely.

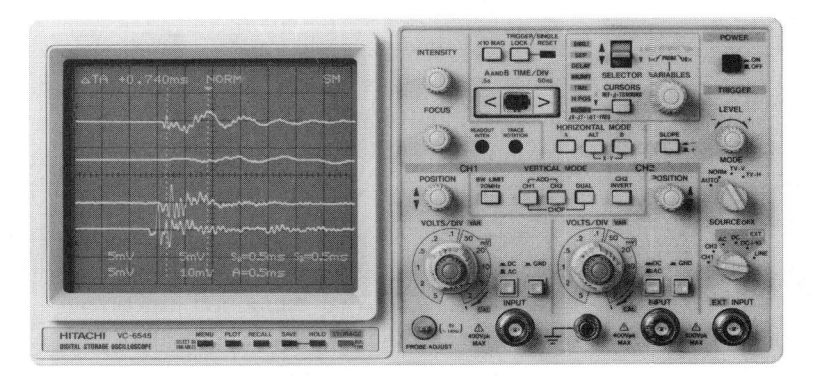

Figure 1.1
An example of the important class of dual-mode oscilloscopes, the model VC-6545 from Hitachi provides a 100MHz bandwidth in real (analog) mode and in digital storage mode, its 40Ms/s sample rate provides a repetitive mode bandwidth of 100MHz. (Reproduced by courtesy of Wayne Kerr, a division of Farnell Instruments Ltd.)

Digital storage oscilloscopes

While oscilloscopes have long been an essential part of the electronic engineer's armoury, their usefulness has been limited in certain applications by the fact that the trace they present on the screen is transitory. When viewing a steady, repetitive waveform this is of no consequence, but many events which require investigation are either unique transients, or if recurring, do so at unpredictable times. An oscilloscope camera provided a means of capturing a trace, for further viewing and investigation at leisure, and the larger manufacturers offered oscilloscope cameras specially designed to fit their products. When a camera was not available, another approach was necessary. One option was an oscilloscope with a long persistence screen. These first appeared early on in the development of the oscilloscope and could often ease problems of viewing a transient. Later still, in the early 1960s, storage oscilloscopes (nowadays generally called analog- or tube-storage oscilloscopes to distinguish them from digital storage oscilloscopes) made their appearance. These enabled a trace to be stored for many hours and viewed for many minutes, though a camera was still necessary if a permanent record was required. Then came the 1970s, and the appearance of the digital storage oscilloscope.

Due to their technical limitations, particularly with regard to sampling speed, early models of digital storage oscilloscopes were viewed as not being in any way in competition with conventional analog oscilloscopes. In fact, they were mainly seen as more flexible replacements for pen- or ultra violet-recorders. The similarity is particularly obvious in *roll mode* of operation, which, like all the other modes, is explained fully in later chapters. In this mode, a digital storage oscilloscope can provide information on what happened just before a triggering event, a facility which has no direct equivalent on a conventional analog scope, even on those using a storage tube.

Since those early days, digital storage oscilloscope design has advanced in many ways. Notably, maximum sampling rates have increased to 1Gsamples/second (1Gs/s) and more and memory lengths have increased to tens or hundreds of ksamples (some models even being able to stream data direct to disk). Furthermore, a host of sophisticated variations on triggering have

Introduction

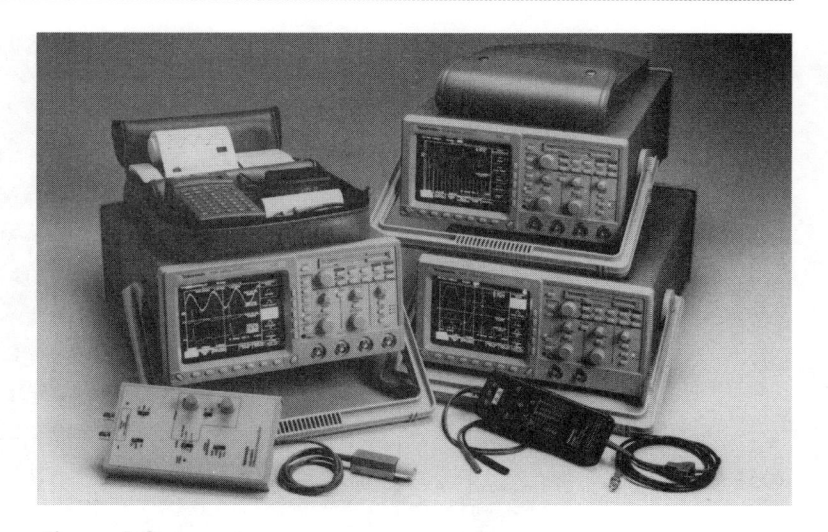

Figure 1.2
Three models from the TDS 4XX range which covers up to 350MHz bandwidth, with 12 bit vertical resolution in hi-res mode. Pictured here with two differential probe systems. On the left, the ADA400A Analog Differential Amplifier providing enhanced sensitivity down to 10mV/division, and on the right the P5200 High Voltage Differential Probe providing safe floating measurements up to 1300V. (Reproduced by courtesy of Tektronix Inc.)

been evolved for special purposes such a glitch capture. The raw sample data can also be processed in various ways, either before or after capture, such as smoothing, averaging etc., while limit-curve windows can be shown on the screen to expedite *go/no go* testing.

Originally large, bulky, heavy and strictly mains operated, digital storage oscilloscopes now come in all shapes and sizes. Portable ones can be anything from little more than a pocket DVM (digital voltmeter) with a facility for capturing and displaying a low frequency waveform, to battery operated digital storage oscilloscopes with an acquisition rate of 500Ms/s or more. The holy grail of digital storage oscilloscope design (which some manufacturers claim to have already achieved) is a digital storage oscilloscope that will do not only all those things that an analog oscilloscope cannot do, but will also do all the things that it can do, and just as well if not better. As will become apparent in the following pages, this is actually a fairly tall order, and so does not come cheap. Some manufacturers have achieved this end by the

3

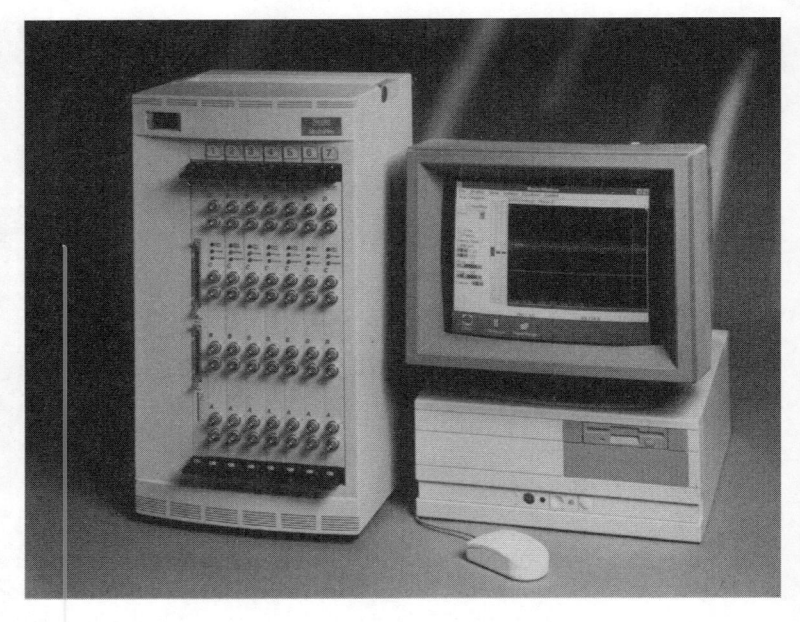

Digital storage oscilloscopes

Figure 1.3
More a digital recording adaptor than a complete digital storage oscilloscope, the Nicolet MultiPro Transient Analyser illustrated features seven, four channel/modules in its transportable tower, and captures up to 28 channels simultaneously. The companion rack-mount version accepts up to 64 channels. Both models provide a 12 bit digitizer and a differential input for each of the four channels on every plug-in unit, and use a PC as the display. (Reproduced by courtesy of Nicolet Instruments Ltd.)

simple expedient of making dual function digital storage/analog oscilloscopes. These are basically analog oscilloscopes with a built-in digital storage facility, the latter using the same display as used in analog mode operation. Admittedly a rather expensive option, it nevertheless provides the user with assurance that what he sees in digital storage mode is real, and not an artifact due to aliasing or whatever, by switching to analog mode and checking that the waveform still looks the same. Providing this same level of confidence in the display in a digital-mode-only oscilloscope requires a considerable degree of ingenuity on the part of the digital storage oscilloscope designer, and no little expense.

Figure 1.4
The Gould model 340 *Easyscope* incorporates a digital multimeter, seven-digit frequency counter and two channel oscilloscope with 20MHz bandwidth. The maximum 20Ms/s sample rate in single shot mode becomes equivalently 400Ms/s in repetitive mode, providing 20 samples per cycle at the full analog bandwidth. Unusually for this class of instrument, the model 340 also functions as an 8 channel logic analyser. (Reproduced by courtesy of Gould Instrument Systems.)

Chapter 2

Digital storage oscilloscope fundamentals

Figure 2.1 is a simplified block diagram of a basic digital storage oscilloscope. Comparing it with the block diagram of an analog

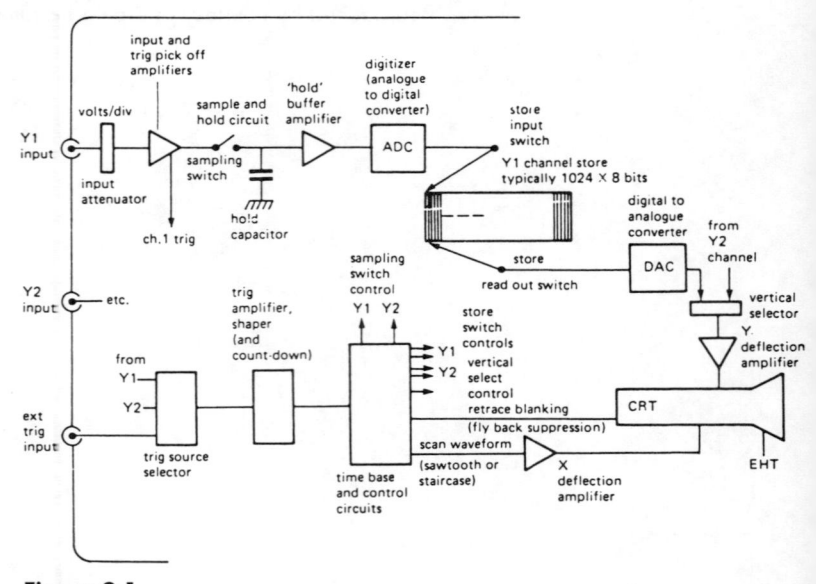

Figure 2.1
Simplified outline block diagram of a typical DSO (digital storage oscilloscope)

oscilloscope would show considerable similarities. The major difference is that the vertical signal, after passing through the input attenuator, Y preamplifier and trigger pick-off stage, is not routed directly to the Y deflection stage. Instead it is sampled at intervals and the samples fed to an ADC (analog to digital converter, sometimes abbreviated to A/D) to be *digitised*, i.e. converted to a string of numbers. Each number represents the voltage at the input at the instant the corresponding sample was taken. The digitised data is stored in a *channel store*, i.e. that part of the total digital memory which is allocated to the particular Y input channel, of which there are usually two and often more. The digital memory consists of a bank of RAM (random access memory) ICs (integrated circuits).

For display purposes, the data currently in the store is read out sequentially and the samples passed to a DAC — a digital to analog converter. There they are reconstituted into a series of discrete voltage levels forming a stepwise approximation to the original waveform. This is fed, along with the reconstituted waveform(s) from the other Y channel(s), to the vertical deflection amplifier to give the usual dual- or four- trace display. Note that the read-out and display of samples constituting the stored waveform need not occur at the same sample rate as was used to *acquire* the waveform in the first place. It is sufficient to use a display sample rate adequate to ensure that each and every trace displayed is rewritten on the screen of the CRT (cathode ray tube) display fifty or more times a second; this is adequate to prevent flicker of the display. This means that in principle the digital storage oscilloscope designer can use a Y deflection amplifier and CRT with very modest bandwidth as the display in a digital storage oscilloscope, even though the instrument as a whole is capable of acquiring and displaying signals with a bandwidth of tens or even hundreds of megahertz. In many instruments, a CRT is not used at all, being replaced by an LCD (liquid crystal display).

In practice, however, a few digital storage oscilloscopes are also capable of being operated as conventional real-time analog oscilloscopes, with a bandwidth in this mode equal to, or even well in excess of, their bandwidth in digital mode. A good example is the Fluke PM3335, with a maximum digitising rate of

Digital storage oscilloscopes

20Ms/s (providing a bandwidth of about 2MHz without sine interpolation) and a real-time analog bandwidth of 50MHz, see Figure 2.2.

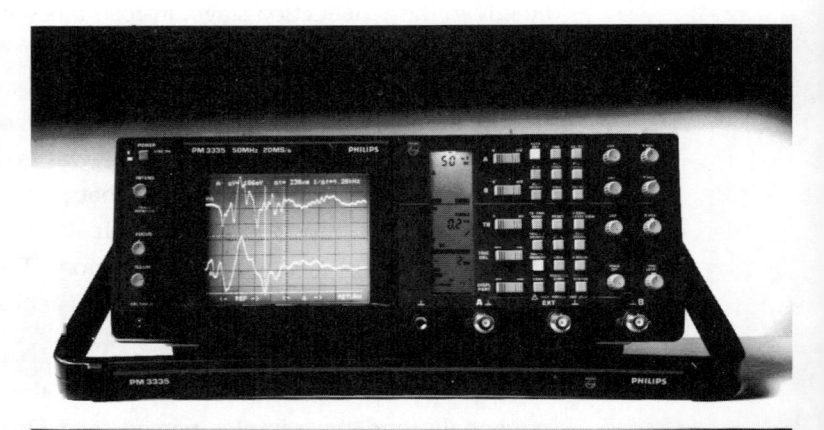

Figure 2.2
The Fluke PM3335 operates both as a conventional real-time analog oscilloscope with a 60MHz bandwidth and as a 20Msample/sec digital storage oscilloscope. In the latter mode, events occuring before a trigger signal can be captured. (Reproduced by courtesy of Fluke UK Ltd.)

There are very real advantages to such a 'dual purpose' instrument, as will become apparent. However, the more common approach is for a manufacturer not to equip a digital storage oscilloscope with a real-time analog capability at all, in which case, all signals displayed are reconstituted from the stored data. In such instruments the display tube may be a raster scanned magnetically deflected CRT, either monochrome or colour — the technology used in TV displays. The digital storage oscilloscopes in the Hewlett-Packard range are good examples of this type of instrument; see for example Figure 2.3. Note that with both the dual purpose and the digital only instruments, however high the sampling rate (and allowing for *equivalent time sampling*, of which more later) the Y bandwidth can never exceed that of the input attenuator and Y input amplifier.

Likewise, however great the vertical resolution (however many

Digital storage oscilloscope fundamentals

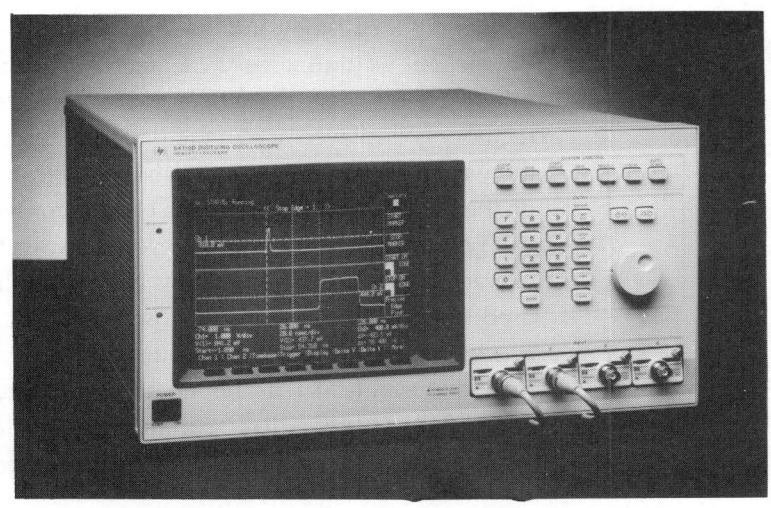

Figure 2.3
Hewlett-Packard Company's colour-display digitizing oscilloscope, the HP 54110D, provides 1GHz bandwidth and a feature set designed for design and test engineers working with high-speed logic and high-speed data communications. (Reproduced by courtesy of Hewlett=Packard Ltd.)

bits the ADC outputs per sample), the vertical measurement accuracy will be limited by the linearity (freedom from distortion) of the Y input amplifier and the ADC. Furthermore, when a dual purpose instrument is used in the analog mode, the horizontal measurement accuracy will be limited by the timebase, X amplifier and CRT linearity to around 2 per cent. By contrast, in the digital storage mode, the measurement accuracy (as distinct from the display accuracy) in the X direction, using cursors, will usually be 0.01% or better.

So much by way of introduction: now for a brief description of the various operating modes of digital storage oscilloscopes, how they work and the implications for the user.

Roll mode

Roll mode operation is rather like a chart recorder, where a trace is written on a strip of paper being drawn at a steady rate from a roll of chartpaper. Imagine the paper to be moving from right to left and you have an analogy of roll mode. The trace on the screen of the oscilloscope appears to be written by a pen hidden just to

Digital storage oscilloscopes

the right of the display and the display scrolls across disappearing off the left of the screen. In fact, in a simple digital storage oscilloscope (where the whole of the trace stored in the display memory is displayed across the screen) information on the part of the waveform off the screen to the left is lost: it does not pile up on the floor like the paper from a chart recorder would. Roll mode can be used simply to keep an eye on a waveform, to observe its characteristics and the limits of its typical amplitude. But it is often used in conjunction with the oscilloscope's trigger circuits, to *freeze* the trace (terminate acquisition) on the occurrence of a particular event, such as the input voltage exceeding some preset limit. A more detailed explanation of how roll mode is implemented will be found in the next chapter.

Refresh mode

When the sample rate (the equivalent of timebase speed in ordinary oscilloscope parlance) becomes too high, the display in roll mode is no longer useful, as the trace appears to rush across the screen and off the left hand side too fast for the eye to be able to make anything of it. A useful alternative in this case is *refresh* or *recurrent mode*: unfortunately the terminology relating to this mode, as with other modes and functions of digital storage oscilloscopes, is not standardised and varies from manufacturer to manufacturer. This mode is particularly useful when the waveform of interest is repetitive or nearly so. With it, the digital storage oscilloscope produces a stable, triggered display looking very like the display on an analog oscilloscope. However, unlike an analog oscilloscope, where the retrace or flyback time is usually only a fraction of the sweep time, in a digital storage oscilloscope it may be much greater than the sweep time, especially in an economically priced instrument where a single microcontroller looks after all of the instrument's functions.

Single shot mode

In this mode, the digital storage oscilloscope is in an inactive state, displaying the last trace captured, until a given sequence of events takes place. The first of these is arming. When the *arm* button is pressed, the oscilloscope commences to acquire data continuously as in roll mode, but without displaying it. When the trigger circuit detects a triggering event, such as the input voltage exceeding the preset threshold set by the user, sampling ceases and the *armed* light goes out. The cessation of acquisition may be either immediate on detection of the trigger event, giving 100% pretrigger information, or delayed by an amount preset by the user, giving some (or all) post-trigger information. The delay may even have been set to correspond to more than the time represented by one screen width. This represents greater than 100% post-trigger delay, and could be used for example to capture, in fine detail for further examination, the last few hundred microseconds of switch-bounce in a relay circuit, which could last for 1ms or longer.

The 100% post-trigger mode corresponds to operation of a conventional analog oscilloscope in the *single shot* or *single sweep* mode, although the analogue oscilloscope needs a camera to record the event, unless it is an analog (tube) storage oscilloscope. However, the ability to capture pretrigger information simply is unique to the digital storage oscilloscope, although it could be done in a rather hit and miss way on a tube storage oscilloscope using *auto refresh* mode.

Equivalent time mode

The sampling rate of a digital storage oscilloscope is limited by the conversion time of the ADC used. To capture a waveform in real time at a timebase speed setting of 100ns/division (with 100 points of display per division) would require an ADC with a sampling rate of 1000Ms/s, or a 1ns conversion time. Such devices are available, but are very expensive. However the resultant bandwidth of a few hundred MHz is relatively easy standard oscilloscope technology for the input attenuator and Y pre-amplifier. *Equivalent time* mode capitalises on this to provide an economical digital storage oscilloscope design employing a less expensive ADC, by arranging for the latter to acquire only, say, ten points on any given timebase sweep. On further sweeps, additional points are acquired, building up a complete picture over many sweeps. How this is arranged is dealt with in detail in the next chapter, but it will be apparent straight away that this mode is only applicable to stable repetitive waveforms.

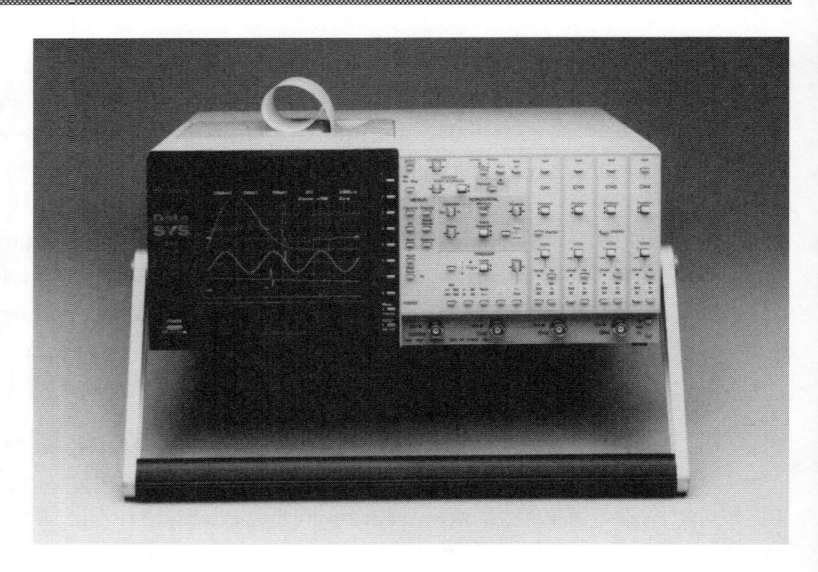

Digital storage oscilloscopes

Figure 2.4
The model 940 four channel digital storage oscilloscope is Gould's top-of-the-range instrument bandwidth-wise, having been upgraded to 400MHz (-3dB). With a 1ns risetime on all sensitivities from 2mV/division up, the maximum sample rate is 100Ms/s or equivalently 5Gs/s in repetitive mode. The wide range of accessories available includes Active (FET), and Differential Probes, and an internal plotter. (Reproduced by courtesy of Gould Instrument Systems.)

Chapter 3

Digital storage oscilloscope modes of operation

Some modes of operation of a digital storage oscilloscope, such as *refresh*, are rather similar to operating modes of an analog oscilloscope, while others, such as *roll* mode have no corresponding counterpart. There is much competition between designers and manufacturers of digital storage oscilloscopes to produce an instrument which handles like, and gives a display virtually indistinguishable from, an analog oscilloscope. Some manufacturers claim indeed to have reached this blessed state, and certainly some of the latest generation of digital storage oscilloscopes are satisfactory for the great majority of applications, although many others — especially the lower priced models — can easily give misleading results in unskilled hands.

One of the modes of operation of a digital storage oscilloscope which have no counterpart in an analog oscilloscope is roll mode, and this is the first mode to be described in this chapter devoted to digital storage oscilloscope modes of operation.

Roll mode

Roll mode is described first, not because it is the most useful mode but because it has been available on digital storage oscilloscopes from the earliest stage, because it is fundamentally different from a conventional 'scope display and because it leads in nicely to the other operating modes of digital storage oscilloscopes. For simplicity, consider a digital storage oscilloscope with 1024 points of memory per input channel, typical of the more economical range of instruments. Some digital storage oscilloscopes display the 1024 points across the usual ten graticule divisions, while others overscan by 2.4%, giving exactly 100 points per graticule division — to simplify the numbers in the following explanations, the latter is assumed here. A fictional engineering application is used to illustrate ROLL mode operation.

Figure 3.1 shows an indeterminate waveform which could correspond to any physical variable — it might be, for example, the voltage output of a load-bearing transducer measuring the stress at one point of a bridge as the traffic passes over. Assume that the digital storage oscilloscope is set to take 100 samples per second, then after (just over) 10 seconds it will have filled up the 1024 memory locations — which are numbered 0 to 1023 — as at A in the illustration. Ten milliseconds later it will be time to take another sample. But before doing so, the digital representations of the samples currently in store in locations 0 to 1023 are read out one after the other and passed in turn to the DAC which turns them back into voltage levels. These are displayed sequentially across the screen from left to right, thus displaying the first ten second segment of the waveform.

Another sample is now taken — but locations 0 to 1023 are already full and there is no storage location 1024. So this new sample is stored in location 0, overwriting the digital value previously stored there. This new 'sample 0' corresponds to a point in time about ten seconds later than the previous sample 0, as at B in Figure 2.4. The channel memory is thus cyclic; like a loop of recording tape, earlier information is replaced continuously by later, as indicated in Figure 3.1. As soon as the new sample is stored in location 0, all the stored samples are cycled through the DAC and displayed again, this time starting with location 1 at the left of the screen, continuing through to location 1023 and finishing up with location 0 (the last sample taken) at the right of the screen. The trace thus displayed is therefore the same as previously, but shunted to the left by one sample position as at B, or by two as at C, etc. Each time, 1024

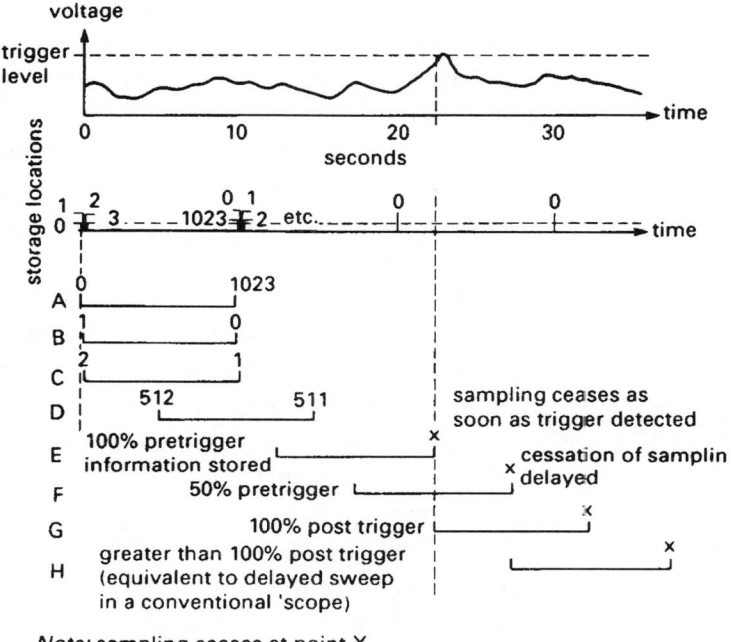

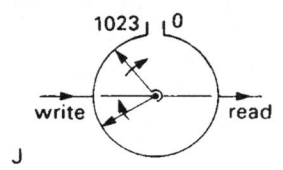

Digital storage oscilloscope modes of operation

Figure 3.1
Roll mode

samples are displayed, finishing with the last one taken, this newest sample being displayed at the right of the screen, the previous earliest sample at the left of the screen being lost. After 512 samples, the original right-hand edge of the trace will have walked across to the middle of the screen, while the left-hand half of the original trace will have disappeared for ever. Alternatively, a display *window* 10.24 seconds long can be considered as advancing along the waveform (see D in Figure 3.1). The important point to bear in mind is that each time the trace is written to the screen of the CRT the samples stored in all 1024 locations are displayed, starting at the left of the screen with the oldest sample and finishing at the right with the sample just taken.

15

Figure 3.1 shows at J the channel memory redrawn as a circular track, with the store input (write) switch and store readout switch of Figure 2.1 drawn like the hands of a clock. In ROLL mode, the write switch rotates continuously at a constant speed. In the example just described, the read switch does not: it rotates once after each sample is taken. If the sample is stored in location n, the read switch reads out the 1024 sample values starting at location n+1 and finishing at location n. It then pauses until the next sample has been taken before rotating once again. Each 'hand' would typically be an eight bit wide data bus: in the case of a memory consisting of dual port RAM this would be a good analogy. However, dual port RAM is expensive and in practice ordinary CMOS or ECL static RAM, or in lower priced instruments, dynamic RAM, is used instead. This has a single read/write data port, which is switched between the two functions by an R/W control line. With the slow data rate in the example just given, there would be no difficulty at all in interleaving the write and read functions, even with relatively slow, cheap, dynamic RAM. If the sample rate were much lower than the 100s/s considered above, the screen trace is rewritten twice or more between each sample, to maintain a high enough screen refresh rate to avoid flicker. On the other hand, at much higher sample rates, several or many samples would be taken before rewriting to the screen.

Returning to the example in Figure 3.1, it is clear that at any time there is a record of the last ten seconds of the waveform in store. This information can be frozen at any time if an event of particular interest occurs — such as a dangerously high stress in the bridge due to an overloaded lorry in the fictional engineering application. The digital storage oscilloscope's trigger circuitry can be set so that if the Y input voltage exceeds a certain level, the sampling action is halted — the write hand in Figure 3.1 ceases to rotate. Furthermore, although the read hand continues to rotate thus continuously displaying the stored trace on the screen, as the trace always starts at the left hand side of the screen with the oldest last sample taken the trace is now stationary. Like the flight data recorder in a crashed plane, the trigger event has terminated the recording of data, rather than initiating it like the trigger circuitry in a conventional analog oscilloscope. Imagine that the flight data recorder uses a loop of tape holding data on just the last ten minutes of flying time, and the analogy is perfect. This type of operation is known as 100% pretrigger store, and is illustrated by E. In the flight data recorder example, the trigger event was effectively the end of the world, but in a digital storage oscilloscope it can be arranged that the circuitry takes some additional samples after the trigger event before terminating the sampling process. Another 512 samples, as point F, will lose the oldest 50% of the pretrigger information but store the first five seconds of the waveform occurring after the trigger event. By suitable settings of the controls, one can in principle have any split one wants between pre- and post-trigger information, or set an even greater delay in terminating sampling, as E to H. In practice, digital storage oscilloscopes usually offer the choice of a small number of settings such as 100 percent, 75, 50 or 25 per cent pretrigger storage, while only the more expensive instruments provide delayed (greater than 100 per cent post-trigger) storage.

Single shot mode

It may be necessary to capture an event which triggers the oscilloscope, but with much greater time resolution than provided by the 100s/s sampling rate in the roll mode example above. But at, say, 100ks/s, the waveform would be rushing across the screen so fast as to present a meaningless jumble to the observer. In this case, triggered storage mode, commonly known as *single shot* or *single sweep* mode, is more appropriate. The digital storage oscilloscope operates in exactly the same way as in roll mode except that the waveform being acquired is not displayed until the trigger event stops the sampling clock — until then, the last waveform acquired continues to be displayed. An *armed* indicator is often provided — this is illuminated while the oscilloscope is continuously acquiring the input signal. When a trigger event occurs, the armed light goes out and the sampling clock is stopped, either immediately (100% pretrigger storage) or when the predetermined desired amount of post-trigger trace has been captured — a *stored* indicator (if provided) then lights. Pressing the arm button releases the lock and rearms the instrument, i.e. resets it into sampling mode, ready to stop upon the occurrence of another trigger event. Operation is very similar

Digital storage oscilloscopes

to that of the single shot mode of a conventional oscilloscope with camera or an analog (tube) storage oscilloscope, with the big difference that with these one cannot capture pretrigger information.

Refresh mode

It was mentioned in the previous chapter that when the sample rate (the equivalent of timebase speed in ordinary oscilloscope parlance) becomes too high, the display in roll mode is no longer useful, as the trace appears to rush across the screen and off the left hand side too fast for the eye to be able to make anything of it. In this case, if the waveform is a stable recurring one, refreshed or recurrent mode is appropriate. With it, the digital storage oscilloscope produces a stable, triggered display looking very like the display on an analog oscilloscope. However, there are subtle differences. In *refresh* mode, when *auto* trigger is selected, the digital storage oscilloscope behaves rather like its analog counterpart. It produces a locked stable picture when triggered, and an unlocked display with successive traces occurring at random timing when not triggered. But a digital storage oscilloscope behaves differently from an analog oscilloscope when they are set to *normal* trigger mode. The analog oscilloscope will produce a locked picture if triggered, but if the *trig level* control is inappropriately set or *external* trigger selected with no external trigger input signal present, the timebase does not run and spot rests blanked at the left hand side of the screen, ready to trace out the waveform if and when a trigger signal arrives. In these circumstances, the digital storage oscilloscope, by contrast, continues to display the last trace acquired.

Like their analog counterparts, digital storage oscilloscopes (when operating in refresh mode) generally ignore anything occurring between the completion of the trace and the arrival of the next trigger pulse. This need not necessarily be so where, as is usual, the digital storage oscilloscope is equipped with separate *acquisition* and *display* memories: acquisition could be organised as follows. Assume that — as is usually the case — the screen display does not correspond exactly to an integral number of

Digital storage oscilloscope modes of operation

cycles of the input waveform. Such a case is illustrated in Figure 3.2, where the screen is shown as displaying about one and a half cycles. So here, half a cycle or so is not being displayed each time the trace is written to the screen. However, the screen display is written using the data stored in the display memory, which is loaded with the same data as the acquisition memory — see point a in Figure 3.2 (a) and (b). (In (b), the waveform has been drawn in a spiral form, indicating continuous writing of the ADC output to the acquisition memory, which is cyclical, as illustrated in Figure 3.1, point J.) The trigger event at point b is ignored, as the display memory address counter has not yet reached 1023. At point C, the ADC output starts to overwrite the data relating to point a on the waveform.

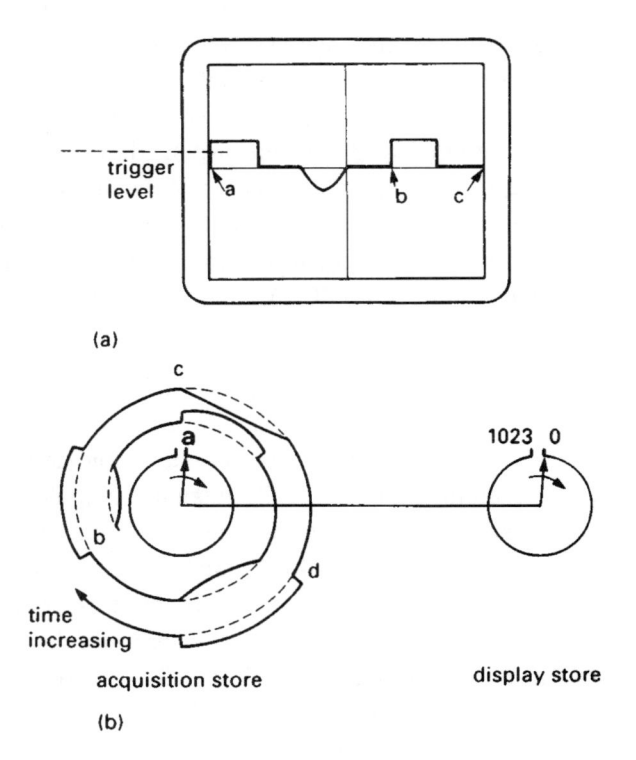

Figure 3.2
Refresh mode

The 'readout switch' copying acquisition data to the display memory continues to 'rotate' in sympathy with the acquisition store write switch, indeed (except in the case of dual port RAM) they are the same thing — the acquisition memory read/write bus. (The acquisition memory read and write 'switches' can in fact be in different positions, or even 'rotate' at different speeds as described in the earlier section on roll mode, simply by supplying different memory addresses (which may also be incremented at different rates) depending upon whether the read/write line is at logic 1 (high) for a read cycle, or logic 0 (low) for a write cycle.)

At point D a trigger event occurs and the acquired data again starts being stored in the display memory as well as the acquisition memory, thus refreshing the stored waveform. However, as the display store address counter stopped clocking up after filling location 1023 on the last 'sweep', the 1024 samples in the acquisition memory starting at the location corresponding to point d on the waveform are stored in positions 0 to 1023 in the display memory. Thus although the segment of waveform c to d was not displayed, it was acquired. A separate trigger at a higher level may have been set to stop acquisition on, or shortly following, a positive- or negative-going large spike or glitch. If such a glitch occurred in an undisplayed portion of the waveform such as C or D, it would duly appear on the frozen display when the last acquisition was transferred to the display memory. (Glitch capture is an important topic covered in detail in a later chapter.)

This arrangement could provide stable triggered viewing of a recurrent waveform while retaining the latest acquisition of that portion of the waveform (between C and D in Figure 3.1) which was not displayed. But in fact, while feasible, few if any digital storage oscilloscopes appear to offer this facility, any information relating to that part of the waveform not displayed on the screen in refreshed mode being lost. This is more important than might appear at first sight, for the following reason. The non-displayed part of the waveform may be likened to that occurring during the retrace or flyback time in an analog oscilloscope. In the latter, the flyback time typically amounts to only a fraction of the trace time at most (although it can be deliberately extended with *hold-off*), while in a digital storage oscilloscope the dead time between sweeps may be as great as or even very much greater than the sweep time, especially if the instrument uses a not very powerful microcontroller. This is especially true of lower priced instruments, where a single microcontroller may be kept very busy looking after all of the instrument's functions — organising the display, scanning the front panel for changes to control settings and possibly doing some pre-processing of the ADC output such as when *averaging* is selected.

While refreshed or recurrent mode is useful for viewing waveforms which include high frequency components and hence cannot be satisfactorily viewed in roll mode, there is a limit to how short a time/division setting it can support while capturing the waveform continuously in the way that has been considered so far. Consider for example a digital storage oscilloscope with an ADC which takes 100ns to convert a sample of the input waveform to the corresponding digital representation, limiting the sample rate to 10Ms/s. Assume the acquisition and display memories each have 1024 points with 2.4% overscan as in the earlier example. Then with 1000 points for the ten horizontal graticule divisions, there will be 100 display points per division and with 100ns minimum per point, the fastest available display speed will be 10ms per division. Depending upon how the points are displayed on the screen (as separate dots, joined by straight lines, or by a *sine interpolator*, of which more later), this will enable the display of waveforms up to, say, 3MHz at most. This is described as the digital storage oscilloscope's *single shot* or *real time* bandwidth. But the bandwidth of the components preceding the ADC — the input attenuator, the Y preamplifier and the sample and hold — will normally greatly exceed this. Given that the input signal is repetitive, it is possible to capitalize on this and produce a much greater effective bandwidth. In this *random repetitive sampling* or *random interleaved sampling* mode, the digital storage oscilloscope does not capture complete chunks of waveform in real time as in Figure 3.2; the waveform is acquired in parts at successive acquisitions, or in *equivalent time*, resulting in a considerably enhanced effective bandwidth.

Equivalent time mode

Continuing with the previous example of a 10Ms/s ADC, imagine that a 100ns/div timebase speed has been selected. With its 100ns conversion time, the ADC can take at most 10Ms/s, so it can only

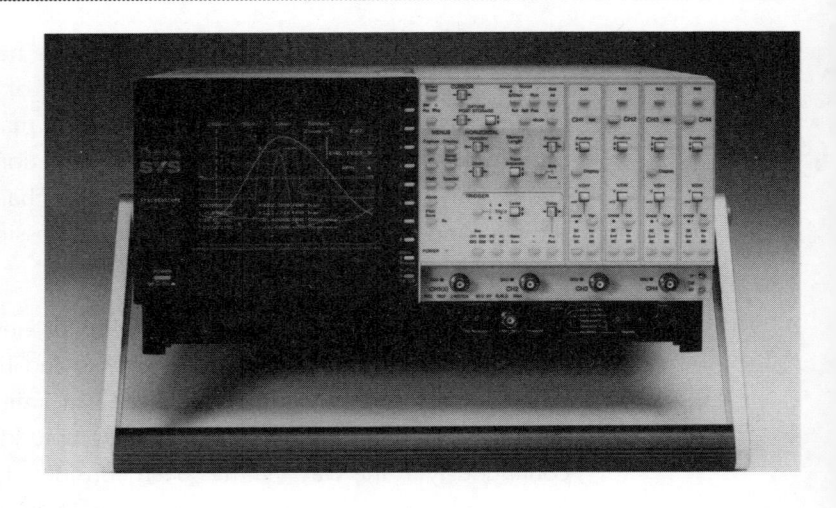

Digital storage oscilloscopes

Figure 3.3
The Gould model 770 Synchroscope is a good example of the increasing number of digital storage oscilloscopes designed with special facilities for a particular area of application. The model 770 is a portable test system for research and test analysis on rotating machines including engines, pumps, turbines and generators. It provides a live, stable display of waveforms and measurements irrespective of rpm. In addition, it is a general purpose four channel 150MHz bandwidth oscilloscope with 10ns glitch detect and numerous other features. (Reproduced by courtesy of Gould Instrument Systems.)

take one sample per division, whereas the display requires 100 samples per division as previously. Furthermore, there will of course usually be no exact relation between the frequency of the input waveform and that of the sampling clock. So the output from the trigger circuit might be just in time to catch the next sampling pulse, or might just miss it, or might occur midway between two samples. Accordingly, the sample in the leftmost tenth of the screen ought to be displayed at the left, right or centre of that horizontal graticule division respectively.

Now at 100ns per division, 100 points per division corresponds to a time per point of just 1ns. Conceptually, the sampling clock for the acquisition memory must run at 1GHz so that the write switch sweeps round all 1024 locations in (just over) 1ms, but as the ADC is only taking 10Ms/s, it deposits a digital sample in only every 100th memory location, as indicated in Figure 3.5 (a), i.e.

Digital storage oscilloscope modes of operation

one per horizontal screen division. However, will it be in the first, second, fiftieth or ninety-ninth location of each horizontal division? The answer is also the solution to the requirement for a 1GHz clock, rather an impractical requirement in a modestly priced instrument, but then it was only a conceptual 1GHz. (Nevertheless, digital storage oscilloscopes frequently describe themselves as having an equivalent sample rate of however many gigasamples per second or GHz effective sample rate.)

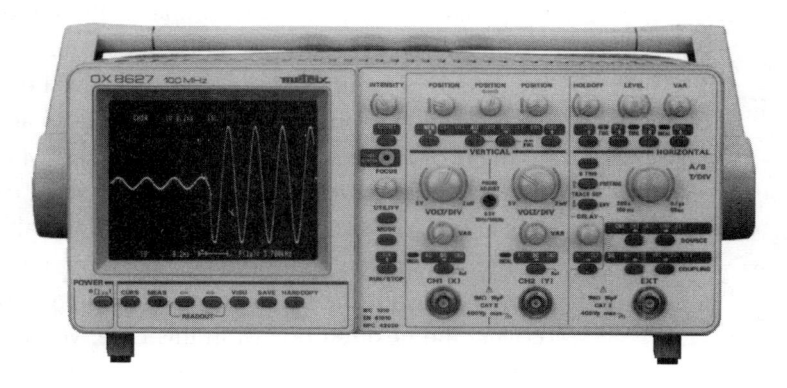

Figure 3.4
The METRIX model 8627 is a dual mode oscilloscope. In analog operation it provides a 100MHz bandwidth with a 3.5ns risetime, from 2mV to 5V/division. Digital storage mode offers up to 40Ms/s, with single shot, roll, refresh, envelope and glitch capture modes.
(Reproduced by courtesy of METRIX S.A.)

The 10MHz sample clock no longer loads the samples taken into successive storage locations in the acquisition store, but into every hundredth location. The trigger pulse starts a timer which measures the delay before the next 10MHz clock pulse arrives. The timer is started by the trigger pulse and stopped by the next clockpulse. The number that the counter has reached is taken as the base address for that acquisition sweep. The first sample is loaded at the base address n , and successive samples at n+100, n+200 etc. Thus each of the ten samples following a trigger pulse is distributed evenly across the acquisition store and hence across the display store and screen, but correctly positioned according to how soon after the trigger the first of the group of ten samples occurred.

With no exact relation between the sample clock frequency and

23

the frequency of the input signal, it is most unlikely that the next trigger pulse will precede a sample clock pulse by exactly the same amount as last time, so now another ten points will be deposited in the appropriate screen positions, adding a bit more definition to the waveform in store, and so on for each succeeding trigger pulse, Figure 3.5. As more and more of the complete picture is built up, it becomes more and more likely that a group of ten points will duplicate an earlier set, so that after 100 trigger pulses the acquisition will still not be complete. But a group of ten points is acquired for every trigger pulse so in just a few milliseconds, thousands of groups of ten points will have been acquired and the picture will be complete. This is so fast as to appear instantaneous to the eye. The exception to this is the case where one is using the 100ns/division sweep speed in order to see a narrow pulse which has in fact a low repetition rate. In this case, although it only takes 1μs to acquire ten points on the waveform, the oscilloscope has to wait awhile for the arrival of the next trigger pulse, when it will collect the next ten points. In this case, you may actually see the waveform picture building up before your eyes.

The equivalent time mode of operation just described is also called multiple point random sampling. It is not unlike the random sampling mode of digital sampling oscilloscopes described in Chapter 9, except that in those, only one sample can be taken per trigger (*sequential sampling*), rather than multiple samples. The reason for this is that the way digital sampling oscilloscopes (which have replaced their analog counterparts, though many of those are still in use) achieve their phenomenal bandwidth is by applying the input signal directly to a super-high-speed sampling gate. This gate is open for such a short *aperture time* that the following hold circuit only has time to charge up to a few percent of the final value. The acquired voltage is then magnified to the full final value and fed back to the floating sampling gate, to bring it to the actual signal value, ready for the next sample.

It is clear how, by using equivalent time sampling, a digital storage oscilloscope operating in random repetitive sampling mode can offer a bandwidth much higher than the frequency of the sample clock, limited ultimately by the bandwidth of the Y preconditioning stages — the attenuator, input preamplifier and

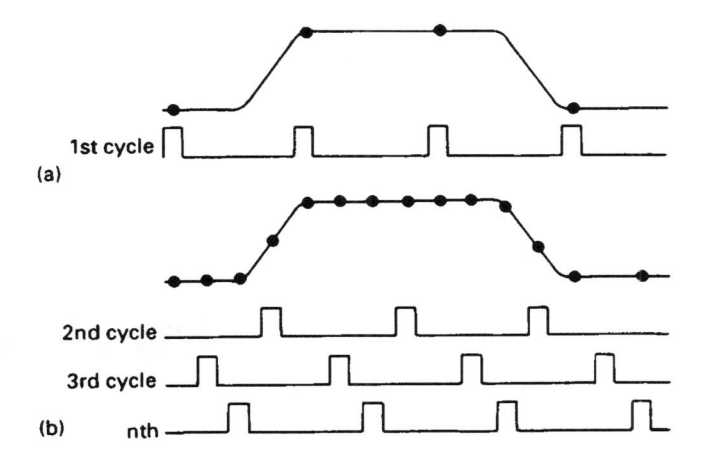

Digital storage oscilloscope modes of operation

Figure 3.5
Multiple point random sampling acquires several points in one acquisition cycle, thus reducing the acquisition time considerably. In this mode a typical DSO would acquire a minimum of 10 points per cycle, so it would reduce the acquisition time by at least an order of magnitude over a scope that acquires a single point on each cycle. (Reproduced by courtesy of Tektronix Inc.)

the sample-and-hold (S/H). The bandwidth of these is sometimes quoted as the *analog bandwidth* in a digital storage oscilloscope specification, even where the instrument uses a raster scanned CRT and consequently can display stored waveforms only — i.e. has no real analog oscilloscope mode.

Where a high real-time (single shot) bandwidth is required, equivalent time sampling cannot be used. The obvious way forward is to use a faster S/H and ADC, topics which are discussed further in Chapter 5.

Chapter 4

Input circuitry

The input circuitry of any oscilloscope, whether an analog- or a digital storage- type, is critical to overall performance. However, the design principles are well understood and design practice long since stabilised, with the result that the input circuitry of a digital storage oscilloscope is identical to that of an analog scope with the same effective bandwidth. An oscilloscope is usually used to examine voltage waveforms. These may exist in a low impedance circuit, such as a power supply or the output stage on a hi-fi amplifier, but they may also be encountered in high impedance circuitry, such as the early stages of a hi-fi amplifier. In the former case the input impedance of an oscilloscope would be of little consequence, but to avoid loading high impedance circuit nodes, and hence changing the amplitude and even the shape of the waveform of interest, oscilloscopes are always designed to have a high input impedance. Regardless of the make or model, the figure invariably chosen is 1M; in parallel with this is some inevitable input capacitance, the value of which is not standardised, varying from model to model, but is usually around 20pF.

The Y amplifier of an oscilloscope traditionally runs at a fixed value of gain, equal to that which it provides on the most sensitive range. For the less sensitive ranges, the input signal is attenuated to bring it down to the same level as on the most sensitive range. Normally, wideband unbalanced variable attenuators are

Input circuitry

designed with a low, purely resistive, characteristic impedance, such as 50 or 75Ω. However, as previously stated, for oscilloscope work, a high input impedance (especially at low frequencies) is generally required, hence the standard value of 1M. At this impedance level, the stray capacitance associated with the attenuator resistors, wiring, switch etc. must be taken into account if the attenuation is to remain constant over a bandwidth of even a few hundred kilohertz, let alone hundreds of megahertz. This is achieved by absorbing the stray capacitance of the components, wiring, switch contacts etc. into larger deliberately introduced capacitances, then adjusting the latter so that the frequency response is constant. The statement that the Y amplifier runs at a fixed gain requires qualification. It will normally run at a fixed gain, but that gain may either be the correct gain to give the rated Y input sensitivity — this is obtained with the variable sensitivity control (VAR) in the CAL position, or the VAR control may be used to reduce the sensitivity by a factor of up to (typically) 2.5 times. This facility fills in between the 1, 2, 5 sequence of sensitivities provided by the Y input attenuator, enabling the display of any amplitude waveform to be adjusted to just fill the screen.

Figure 4.2 shows a typical input attenuator such as might be found in a modest oscilloscope of 20MHz bandwidth. It can be seen that each attenuator pad , e.g. the ÷10 position using R3 (900K) and R4 (effectively 100K owing to R9 plus R10 in parallel with it), has a capacitive divider CV4 and C3 in parallel. CV4 is adjusted so that its value is one ninth of C3 + CV9 + the input capacitance of the Y amplifier. Thus the resistive and capacitive division ratios are the same and the attenuation is independent of frequency. CV9 is used to set the input capacitance of the oscilloscope on the most sensitive range to its design value, while CV1, CV3, CV5 and CV7 enable this same input capacitance to be set on all the other input ranges. This is important when using a passive divider probe, as this is similarly internally compensated. R9, C6 and CV10 protect the input stage of the following Y amplifier from damage in the event of a large input at dc or low frequency being applied to the oscilloscope on its most sensitive range, while passing high frequencies attenuated by 10%, the same as due to R9 and R10. It is therefore important that large amplitude signals at high frequencies, e.g. the output of a

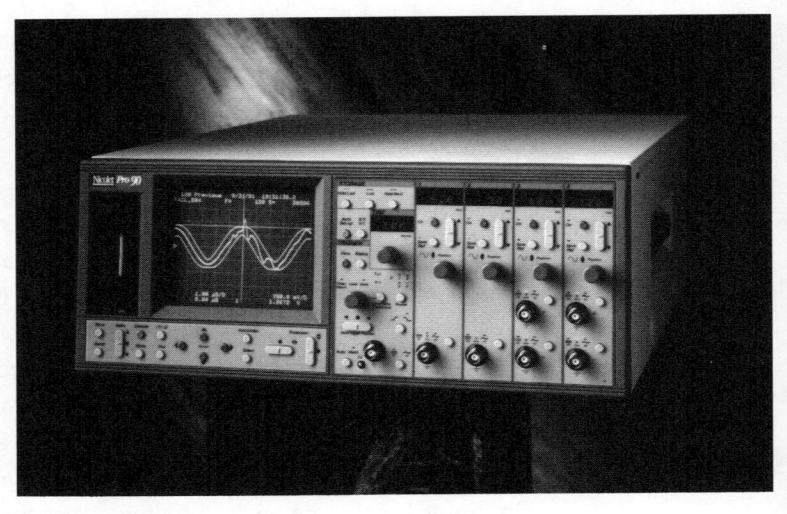

Digital storage oscilloscopes

Figure 4.1
The versatile four channel PRO90 is virtually two digital storage oscilloscopes in one. Two channels feature 8 bit sampling at 200Ms/s max, providing a repetitive bandwidth of d.c. to 100MHz.. Channels 3 and 4 provide 12 bit resolution and differential inputs, furnishing a 5MHz bandwidth (except on the three most sensitive ranges). An internal disk recorder is included, and like most modern oscilloscopes, both analog and digital, it features auto-set-up, selecting suitable input voltage and timebase ranges, and trigger level. (Reproduced by courtesy of Nicolet Instruments Ltd.)

radio transmitter, should not be applied to an oscilloscope on the more sensitive ranges, as damage may result. It is also worth noting that the input impedance of an oscilloscope is not constant. At dc it is 1MΩ, and remains so up to a few hundred hertz. Thereafter, it becomes a predominantly capacitive reactance falling with increasing frequency, being typically only 4K at 10MHz. Provision is also made for selecting dc coupling, ac coupling (with a low frequency −3dB point of 1.6Hz in Figure 4.2, though other oscilloscopes often exhibit a 10Hz −3dB point when AC coupled) or Y amplifier input grounded — useful when adjusting the Y shift to centre the trace vertically.

The Y amplifier stage of an oscilloscope (digital or analog) of high performance, was traditionally designed using discrete devices, the first stage being an FET (field effect transistor) long-tailed pair. This provided a high input impedance combined with low offset drift with change of ambient temperature. However, in the

Input circuitry

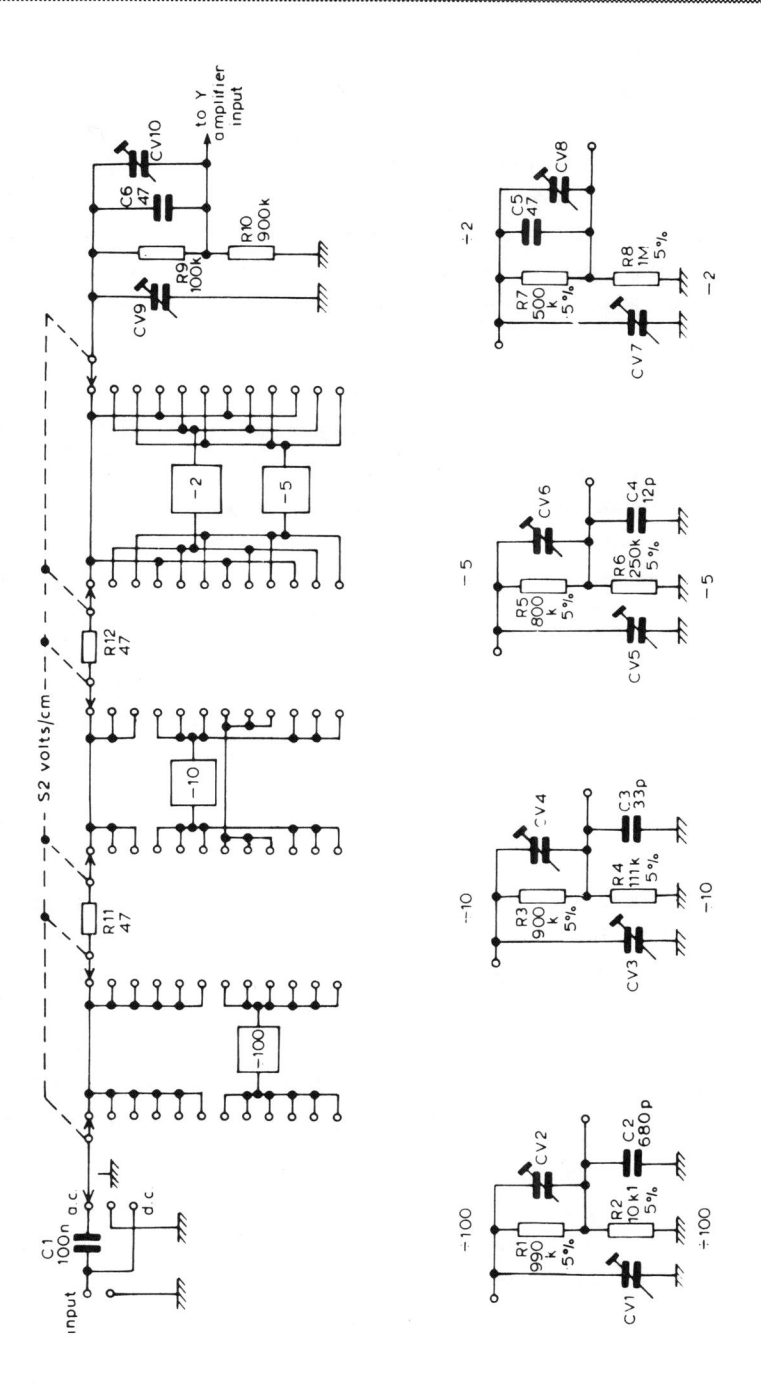

Figure 4.2
Frequency-compensated input attenuator as used in the Scopex 4S6. (Reproduced by courtesy of Scopex Instruments Ltd.)

pursuance of increased sales through offering customers better value, many manufacturers now use ICs extensively, the major manufacturers having in-house semiconductor development and production facilities. A balanced input stage is still used, for low drift, even though most oscilloscopes only offer unbalanced inputs. Full use of the balanced input stage is made in those oscilloscopes (several of which are illustrated in this book) which offer balanced inputs. These oscilloscopes are capable of accepting centre-grounded or floating two-wire input signals.

The Y input amplifier converts the signal from balanced to unbalanced (if the instrument has a balanced input) and amplifies it up to a level suitable to apply to the ADC. It also supplies a copy of this signal, at a suitable level, to the trigger circuitry. There is a noteworthy difference of design philosophy between analog and digital oscilloscopes. In the former, the linear range of the Y amplifier, before distortion sets in, corresponds to two or three times the vertical display window — the range covered by the graticule in the Y direction. This permits the fine detail of, say, the positive peak of the waveform to be examined in more detail by increasing the Y sensitivity, and shifting the trace down with the Y shift control to bring the positive peak back on screen. In a digital storage oscilloscope with a 12 or 14 bit ADC, the same design philosophy may be employed, but in models with only an 8 bit ADC (models have even appeared with as few as 6 bits of vertical resolution), a two- or three-times overscan capability throws away one or one and a half bits of on-screen resolution. So in digital storage oscilloscopes and storage adaptors, vertical overscan capability tends to be minimal. Instead, parts of a waveform can be examined in greater detail, using post-storage expansion in the Y direction, and, on many digital storage oscilloscopes, in the X direction also.

There is another function which in modern oscilloscopes (of either the analog or digital storage variety) is often undertaken by the Y amplifier. In addition to a gain range of 2.5:1 provided by the VAR sensitivity control, the Y amplifier gain may be adjusted by a factor of x1, x2 or x4. These correspond to Y deflection factors of 20, 10 and 5mV/division, and are selected as appropriate by the Y input attenuator switch. Thus the Y input attenuator can be drastically simplified to just x1, x10 and x100 positions, the x2 and x5 compensated attenuator pads of Figure

Input circuitry

Figure 4.3
The autoranging TEKSCOPE combines the functions of a digital multimeter/frequency meter and a 100MHz digital storage oscilloscope. (Reproduced by courtesy of Tektronix Inc.)

4.2 being no longer required.

One potential problem with the display of signals on a digital storage oscilloscope is aliasing. This phenomenon results in a waveform of one frequency apparently displaying a quite different (usually much lower) frequency. There are two quite different forms, *perceptual aliasing* (an artefact of the dot display, and covered in Chapter 9), and *true aliasing*. The latter occurs when the signal of interest is at a frequency comparable with or higher than half the sampling rate, or in the case of a complex waveform, contains significant energy at frequencies near or above half the sampling rate. A digital storage oscilloscope, like any sampled data system, is ignorant of what happens between samples. Provided there are no dramatic changes in level between samples, the reconstructed waveform will be an acceptable approximation to the original, even though it is quantised both in time and in level. Formally, the requirement is that there is no significant signal energy at or above the *Nyquist rate*, i.e. at or above half the sampling rate. Most digital storage oscilloscopes leave it up to users to satisfy themselves that this is the case. One way to ensure that the Nyquist criterion is met, is to include a low pass filter in each Y input channel, ahead of the ADC. As the filter's cut-off frequency needs to be adjusted in sympathy with the X timebase time-per-division switch, this is an expensive option and hence not commonly found. A cheaper alternative is to feed the Y input signal to a frequency counter, which is arranged to light a warning light if the input frequency is near the Nyquist rate at the selected timebase speed. Generally, even with a simple waveform there should be ten samples per cycle. The exception is if the waveform is known to be sinusoidal, in which case a *sine interpolator* circuit, if provided, can be switched in. This enables a sinewave to be reproduced with as few as three or even two and a half samples per cycle. However, such a sine reconstruction filter operates on the stored data at display time, and is not part of the Y input chain.

Chapter 5

Digitising

The input signal from the attenuator and Y preamplifier is digitised by an ADC (analog to digital converter, also written A/D), so that it can be stored in the acquisition memory. Each sample of the signal is converted to a binary number, as digital circuitry does not use the decimal notation with which we are familiar. In the latter, there are ten digits, 0 to 9, and larger numbers are indicated by putting digits in the tens, hundreds or thousands columns etc. as necessary. In binary notation, there are only the two digits 0 and 1, larger numbers being represented by putting 1s in the twos, fours, eights columns etc. Thus the decimal digits 0 to 5 are shown in binary as 0, 1, 10, 11, 100 and 101 respectively. Clearly, for larger numbers, the representation becomes unwieldy, so a more compact form called hexadecimal (*hex*, or *sixteens*) notation is used. Here, the numbers 0 to 15 are represented by the digits 0 to 9, by A for ten, B for eleven and so on up to F for fifteen. A hex digit or four bit binary number is also known as a *nibble*, two nibbles making a *byte*. Thus a byte consists of eight *bits* or binary digits. Now an eight bit binary number can be written with just two characters, representing any decimal number up to 255 (15x16 + 15), written FF_H in hex. Note that the suffix H is always used with hex numbers to avoid confusion with decimal numbers represented by the same digits. Thus 95_H represents the decimal number (9x16 + 5), or one forty nine decimal, 149_D.

Many digital storage oscilloscopes, especially all the lower priced models, use an 8 bit ADC. Models which are more expensive may use 12 bit or 14 bit converters, giving a resolution in the Y direction of one part in 4096 or in 16384 respectively, as against one in 256 for an eight bit converter. Clearly, post storage Y expansion will reveal additional detail in an instrument with 12 or more bits of resolution whereas, in an instrument with an 8 bit ADC, post storage Y expansion is of much more limited use (except where averaging is in use, providing one or more additional bits of resolution). Nevertheless, 8 bit ADCs are commonly found even in more expensive instruments. This is because 8 bit converters are much faster than those with 12 or 14 bits and consequently digital storage oscilloscopes with a high maximum sampling rate, 500Msamples/second or more, use 8 bit ADCs — indeed some early wideband models used a 6 bit ADC.

There are two main types of ADC used in digital storage oscilloscopes, the *successive approximation register* (SAR) type and the *flash* converter. Figure 5.1 shows the block diagram of an SAR ADC. The heart of the system is the successive approximation register, which controls a high speed DAC (digital to analog converter). On receipt of a convert command, the SAR sets the MSB (most significant bit) of the n-bit wide word controlling the DAC to 1 and all the other bits to zero. The output of the comparator will be a 0 or a 1, depending on whether the input is lower or higher than the DAC output, i.e. lower or higher than one half of full-scale input. The SAR latches the comparator output, freezing the DAC MSB output, and sets the NMSB (next most significant bit) to 1. The comparator output will now be a 1 if and only if the input is greater than three quarters of full scale or between one quarter and one half of full scale. Again the comparator output is latched, and successively less significant bits tested in the same way, building up the full n-bit answer, with the control and timing logic signalling the end of conversion as soon as it is ready.

Figure 5.2 shows a block diagram of the other main type of analog to digital converter, the flash ADC. This is equipped with $2^n - 1$ comparators, each with its invert input tied to one of $2^n - 1$ reference voltages obtained from a string of resistors connected to −Vref and +Vref. All of the non-inverting comparator inputs are tied together and the input voltage is applied to them. Consequently, if the input voltage is at 0V (ground potential), all

Digitising

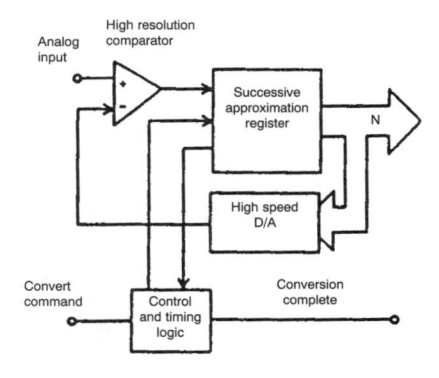

Figure 5.1
A successive approximation analog to digital converter. (Reproduced by courtesy of Analog Devices.)

of the comparators whose inverting inputs are connected to a reference voltage less than this will output a logic 1, while all the rest will output logic 0. The decode logic converts this result to the corresponding n-bit binary number and outputs this to the external world. Clearly, with this type of converter, the digital result representing the input analog voltage is available continuously; no commence conversion command is necessary.

By contrast, it is clear that for an SAR type of ADC to operate correctly, the input signal must not change by even the smallest amount during the course of a conversion. Suppose for example an eight bit SAR ADC covering the range (say, for convenience of explanation) of −1.28V to +1.27V, converting an input at 0V. When the conversion starts, the MSB will set to 1, but if the input then falls to just −10mV, all the remaining bits will also set to 1, giving an answer of FF_H, corresponding to +1.27V, instead of $7F_H$ corresponding to (the now correct answer of) −10mV. So in a digital storage oscilloscope, where changing input voltages are the order of the day, a successive approximation type of analog to digital converter is always preceded by an S/H (sample and hold) circuit. The latter consists essentially of a sampling switch, hold capacitor and hold buffer amplifier, as shown situated between the output of the Y preamplifier and the input of the analog to digital converter in Figure 2.1. The buffer amplifier is designed with a high input resistance, so as not to discharge the hold

Digital storage oscilloscopes

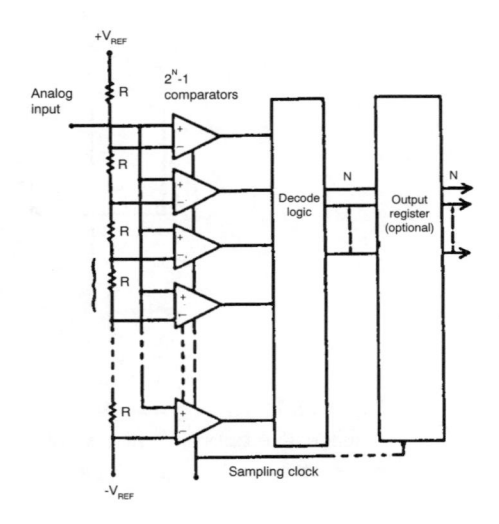

Figure 5.2
A flash analog to digital converter. (Reproduced by courtesy of Analog Devices.)

capacitor during a hold period, when the switch is open. The switch is opened by the leading edge of the start conversion command to the ADC, and closed again as soon as the conversion is complete. Thus the voltage on the hold capacitor is forced to follow the input voltage until *frozen* again at the start of the next sample. This technically makes it a *track and hold* rather than a sample and hold, but when sampling at the maximum possible rate, there is of course no difference. The requirement to acquire the current level of the input voltage virtually instantaneously when switched back from hold to sample, demands that the hold capacitor shall be as small as possible, hence the need for a very high input resistance in the hold buffer. The various timing aspects of sample and hold operation are detailed in Figure 5.3.

Both SAR and flash analog digital converters face a speed/resolution trade-off, and a great deal of ingenuity has been expended to obtain both high speed and high resolution. One method is the sequential partial conversion or sub-ranging approach. Here, the conversion takes place in two parts, the first yielding a k-bit answer and the second an m-bit answer, where k + m n. For example, a four bit flash converter's output could be

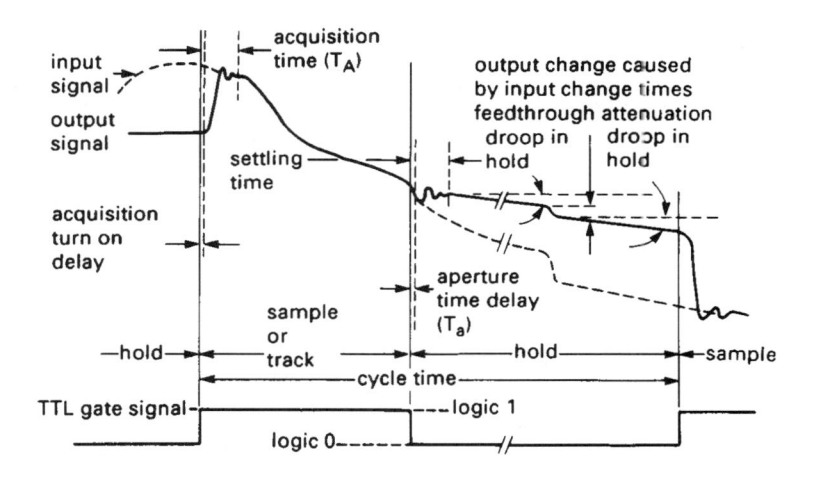

Digitising

Figure 5.3
The elements that make up the acquisition cycle of an ADC. The turn on time or the time the device takes to get ready to acquire a sample is the first event that must happpen. The acquisition time is the next event that occurs. This is the time that the device takes to get to the point at which the output tracks the input sample, after the sample command or clock pulse. The aperture time delay is the next occurrence. This is the time that elapses between the hold command and the point at which the sampling switch is completely open. The device then completes the hold cycle and the next acquisition is taken. (Reproduced by courtesy of Tektronix Inc.)

latched and fed to a four bit DAC (both with better than 8 bit accuracy). The DAC output can be subtracted from the input, the residue amplified by a factor of sixteen then fed to the same or another four bit flash converter, giving a two pass 8-bit result. Figure 5.4 shows a sub-ranging 12-bit converter. Here, as k + m > n, digital correction is included to allow for overlap between the MSB of the 5-bit first flash stage and the LSB of the 8-bit final flash stage. It will be clear that, like SAR analog to digital converters, sub-ranging converters also require a sample and hold stage, so that the pure flash converter is the outright winner in the sampling-speed stakes. However, flash converters with much more than 8 bits are clearly impractical due to the very large number of comparators which would be required.

There remains a topic which is not directly concerned with digitising, but which nevertheless can conveniently be dealt with

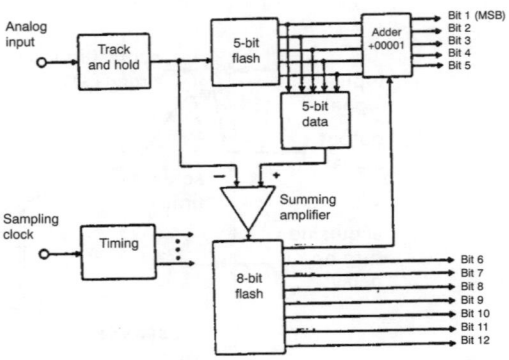

Digital storage oscilloscopes

Figure 5.4
A two pass analog to digital converter. (Reproduced by courtesy of Analog Devices.)

in this chapter. This concerns *charge-coupled devices* (CCDs), which can alleviate the cost/speed trade-off of analog to digital conversion. Charge-coupled devices have been available for many years. They are sampled analog clocked delay lines in which a packet of charge, representing the amplitude of the input voltage at a given instant, is shunted along from one stage to the next at each clock pulse. The string of samples eventually emerges from the output end of the line after a delay equal to the clock period times the number of stages, typically 512 stages. The CCD is an example of a *pulse amplitude modulation* (PAM) device, that is to say the information is quantised in the time domain but not in terms of amplitude. The amplitude of each sample emerging from the line can take any value (within the device's dynamic range), rather than being forced to the nearest discrete level as it is in an ADC. Continued development of these devices has raised their top operating frequency to several hundred Msamples per second. So whereas at one time it was necessary to employ several in parallel with staggered clock drives in order to achieve an effective sample rate of 400Ms/s or more, this can now be achieved in a single device. The beauty of the scheme is that a single shot bandwidth of well over 100MHz (with sine interpolation) can be obtained with a relatively slow analog to digital converter. This is achieved as follows. When a trigger event stores a high speed transient, it does so by stopping the CCD clock. This freezes a string of 512 analog samples in the CCD delay line. A lower frequency clock is now applied so that instead of spilling out of the end of the CCD delay line at up to 500Ms/s, the samples now trickle out at a rate within the capabilities of a fairly modest, inexpensive ADC. In refreshed mode, the ADC

Digitising

converts only every umpteenth sample from the delay line (via an S/H), building up the picture in equivalent time, whenever the CCD clock frequency exceeds the maximum ADC conversion rate. At lower clock rates, such as in roll mode, the ADC can cope with all of the samples out of the CCD delay line as they arrive.

As the performance of high-speed analog to digital converters has advanced and their price fallen, it was forecast that the use of CCDs (which were considered to have been developed to their maximum potential) in digital storage oscilloscopes would become a thing of the past. However, this has not happened, at least not yet, with CCDs holding their own in some models of high sample rate digital storage oscilloscopes. Note however that although the need to use several CCD delay lines harnessed in parallel with staggered clocks, with a multiplexer at the back end to present samples to the ADC in the appropriate order, has been reduced by the development of faster CCD delay lines, staggered clocks are still used in the interests of high sample rates. They are applied to several fast ADCs harnessed in parallel, to achieve much faster effective sampling rates than can be provided by a single ADC. This scheme is also used in some two- and four-channel digital storage oscilloscopes, to provide twice the sample rate, and hence twice the bandwidth, when only one or two channels are in use respectively.

Finally in this chapter, a word about storing the samples from the analog to digital converter in acquisition memory. With ADCs running at up to 500Ms/s, or in the latest models over 1Gs/s, it is clear that even the most expensive, power-hungry *random access memory* (RAM) will not be fast enough to accept all the samples as they arrive from the ADC. The problem seems insoluble, but fortunately it is not, for the samples are stored in a very specific order, so that a true random access at the full rate is not required. A demultiplexer is used to distribute the samples to parallel banks of RAM, so that each bank is only accessed at a lower rate. Figure 5.5 shows the scheme; here, four banks of RAM are shown, permitting a fastest sample clock period of just one quarter of the RAM write cycle time, but the scheme can be extended to eight or sixteen parallel banks as necessary to support the desired sampling rate. A multiplexer is also required, to present the acquired data to the display store in the correct order, but only the demultiplexer following the ADC has to work at the maximum sample rate.

Digital storage oscilloscopes

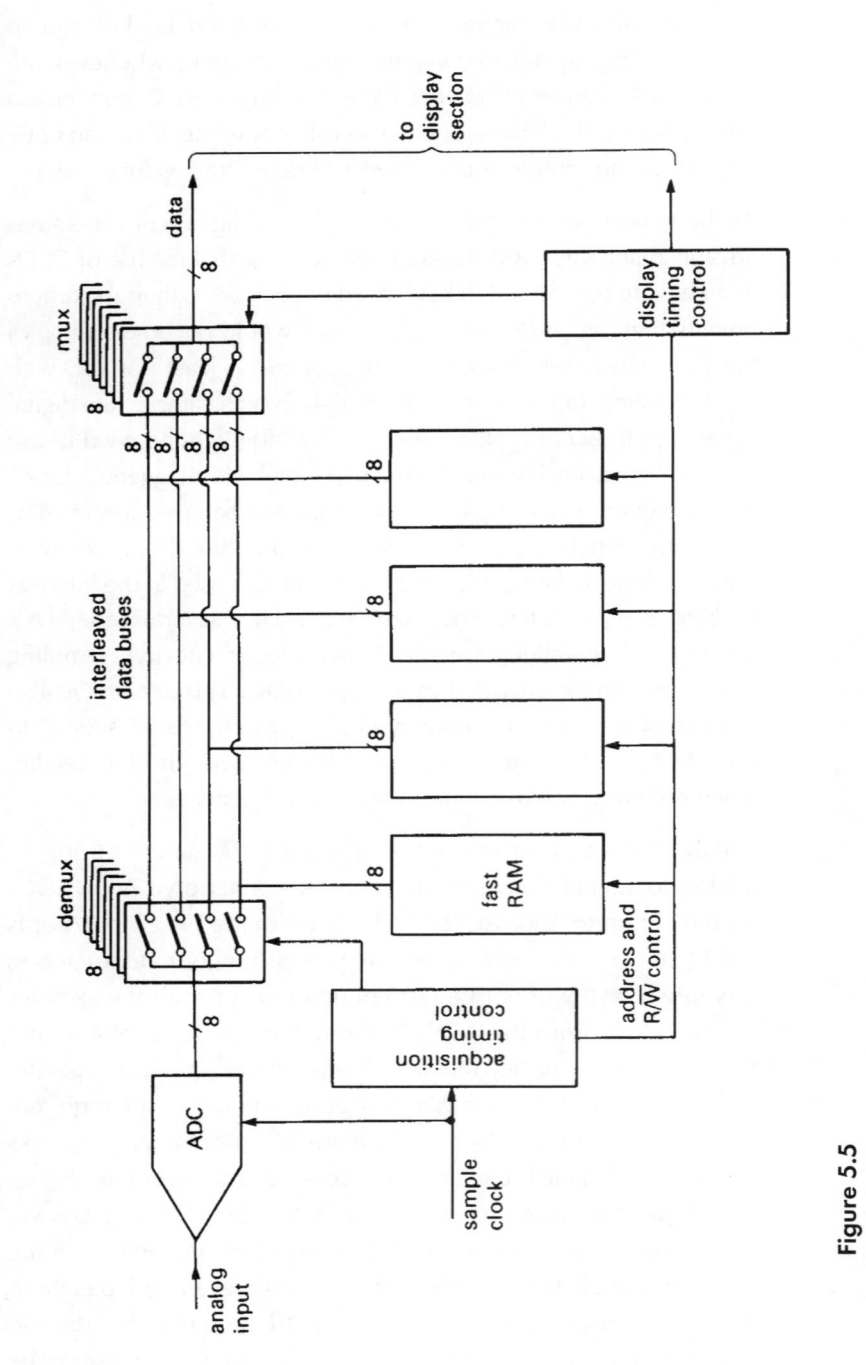

Figure 5.5
Outline schematic showing how the acquisition memory RAM can handle up to four times the data rate of the individual RAM ICs. Further subdivision to eight banks and/or dual port RAM would provide even greater speed.

Digitising

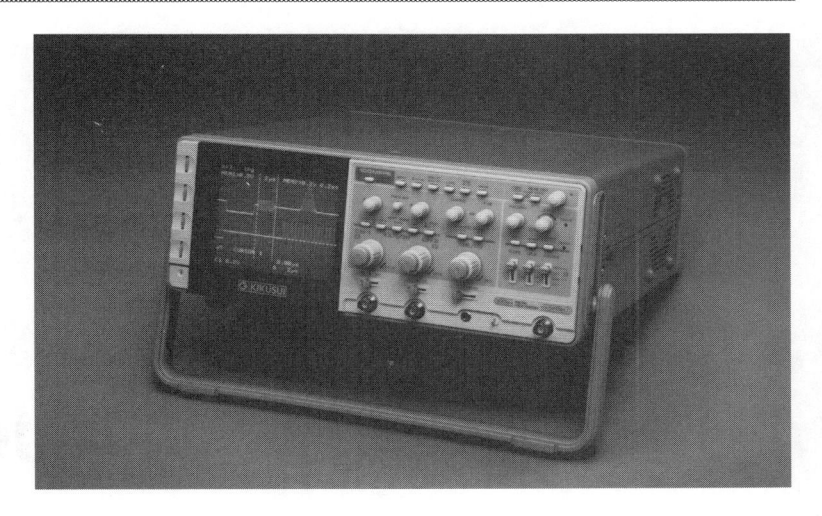

Figure 5.6
Another good example of the important class of dual-mode oscilloscopes, the model COR5502U from Kikusui provides a 100MHz bandwidth in real (analog) mode and in digital storage mode, it has a 26.9MHz single shot bandwidth (with sine interpolation) and a repetitive mode bandwidth of 100MHz. It is the top-of-the-range model in 5500U series of instruments and features timebase speed ranges down to 2ns/division (with x10 magnifier). (Reproduced by courtesy of Teleonic Instruments Ltd.)

Chapter 6

Triggering

Digital storage oscilloscopes can do most of the things that analog oscilloscopes can do, as well as a number of things that they can't. So it is not surprising that many of the controls on the two types of instruments are very similar. This applies to the trigger department as well as other departments. Thus DSOs usually offer a choice of INTERNAL TRIGGER from CHANNEL 1 or CHANNEL 2 (on a two channel instrument, or any of the channels on a four channel instrument), or EXTERNAL TRIGGER. Most instruments offer an EXT÷10 input to increase the dynamic range of acceptable external trigger signals. Additionally, mains-powered instruments provide a built-in LINE TRIGGER (i.e. from the 50 or 60Hz AC line). Also provided are the usual AUTO and NORMAL modes: in the latter, the timebase only runs if a trigger is present, failing which, the last trace acquired continues to be displayed. AUTO causes the timebase to retrigger every 20ms or so if no further triggers are received. A SINGLE SHOT position is also commonly available, arming the instrument to make a single acquisition when a trigger signal arrives, or alternatively freezing acquisition immediately on receipt of a trigger (or after a delay) to

Triggering

provide 100% pre-trigger information (or both pre- and post-trigger signal capture). The TRIGGER coupling may be set to DC, AC, AC HF Reject or AC LF Reject, and additionally some instruments provide a sync separator so that triggering from line or frame sync pulses of a baseband TV signal may be selected. If DC trigger coupling is selected, the trigger point set by the TRIG LEVEL control will correspond to a fixed level on the display, so that if the waveform is moved up or down on the screen with the Y SHIFT control, the point on the waveform at which triggering occurs will alter. It will likewise alter if the DC level of the input signal changes, assuming that the Y INPUT is also set to DC coupling. If any of the AC coupled TRIG modes is selected, then the TRIG LEVEL control will set the position on the waveform at which triggering occurs, independently of the setting of the Y SHIFT control. Naturally, a choice of positive- or negative-going sync is available, as well as a level control to select the level on the waveform at which triggering occurs. With positive trigger selected, the level control will set the trigger point anywhere desired, from a point slightly above the negative peak, right up almost to the positive peak. The exception is when examining fairly low frequencies with the LF Reject facility in use, or fairly high frequencies with HF Reject selected. These controls cause a progressive decrease in trigger circuit sensitivity at low and high frequencies respectively. Besides decreasing the effect of unwanted low- or high-frequency components on triggering from the wanted waveform, these controls have an incidental effect upon triggering, due to the phase shift that they introduce, which is worth noting.

Suppose that HF Reject has been selected, on an oscilloscope where this mode rolls off the high-frequency response of the trigger channel above 50kHz, in order to obtain a stable display of a 50kHz sinewave that has superimposed on it some narrow spikes of an unrelated frequency. The trigger circuit will reject the spikes and respond only to the wanted 50kHz sinewave, which will thus be cleanly locked, although the spikes may be visible running across the trace, unsynchronised. If very narrow, they may well be quite invisible at the timebase speed used to view the 50kHz waveform, yet without the HF Reject facility they could have made it quite impossible to obtain a locked picture of the wanted signal. If the digital storage oscilloscope in use had a MAX/MIN or ENVELOPE mode which runs the ADC at its maximum conversion rate whatever the timebase speed, then the spikes would be visible, especially if the INFINITE PERSISTENCE mode (where any signal occurring is stored indefinitely on screen) — if provided — were in operation. These are facilities found on many digital storage oscilloscopes which have no parallel in an ordinary analog oscilloscope. However, the effect of the phase shift introduced by the LF and HF Reject controls still applies.

In the foregoing example, the trigger circuit will respond to the

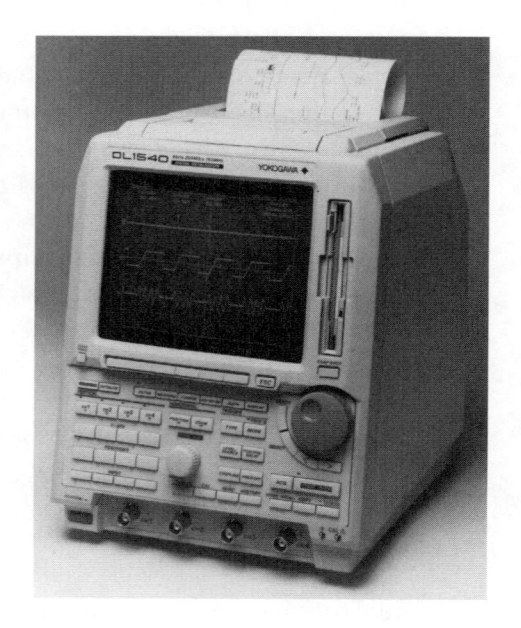

Figure 6.1
The four channel DL1540 features 8 bit sampling to 100Ms/s, providing a single shot bandwidth of 40MHz (200Ms/s, 80MHz in two-channel use) and a repetitive bandwidth of d.c. to 150MHz. A zoom facility makes good use of the vertical resolution, can be increased to 9 bits with digital filter or 12 bits with averaging. Other standard features include 56kword per channel memory and both Zone and Parameter GO/NO-GO comparison functions, whilst options include a built-in printer. (Reproduced by courtesy of Martron Instruments Ltd.)

wanted 50kHz sinewave, although its response will be 3dB down, i.e. the smallest 50kHz sinewave that it wil lock on is about 40 per cent larger than at much lower frequencies (assuming that the HF Reject filter is a simple single pole type, as is usual). In addition, there will be a corresponding 45° phase lag in the trigger channel. The significance of this is that if manual trigger, positive-going has been selected the TRIG LEVEL control will no longer permit triggering at any desired level on the positive-going flank of the sinewave. Instead, the TRIG LEVEL control will initiate the sweep anywhere from (just above) one quarter of the way up the positive flank, to almost one quarter of the way down the following negative-going flank. At frequencies higher than this the effect will become even more pronounced. A similar effect will be noticed when triggering from a sinewave with a frequency near the LF cut-off frequency when LF Reject is selected, except that in this case there will be a phase advance, so the trigger range will be advanced by up to a quarter of a cycle or even more, rather than retarded as in the HF Reject case. Different makes and models provide various -3dB *corner* or *breakpoint* frequencies for their trigger circuits. For example, on the Nicolet PRO series of instruments, the LF Reject corner frequency is 10kHz, while the HF Reject is set at 1MHz, whereas some other makes set the HF Reject frequency as low as 15kHz.

Trigger circuitry in digital storage oscilloscopes (especially those designed with the accent on logic analysis) often offers more functionality than that found in straightforward analog oscilloscopes. Figure 6.2 (a) shows the effect of window triggering, which is useful for catching glitches or overvoltage conditions. Usually, each level is independently settable by the user. Figure 6.2 (b) shows hysteresis triggering, which makes the trigger point less susceptible to noise. It allows a level and slope trigger to occur only after the signal has crossed a hysteresis level. This acts as a trigger enable. It is clear from Figure 6.2 (a) that a glitch which remained within the peak to peak excursion of the waveform could not be detected by simple window triggering. The Gould DataSYS series of digital storage oscilloscopes feature *V-Glitch* trigger, which is a true dV/dt function, designed to trigger the oscilloscope if the signal change between any two acquisition points is greater than a set amount.

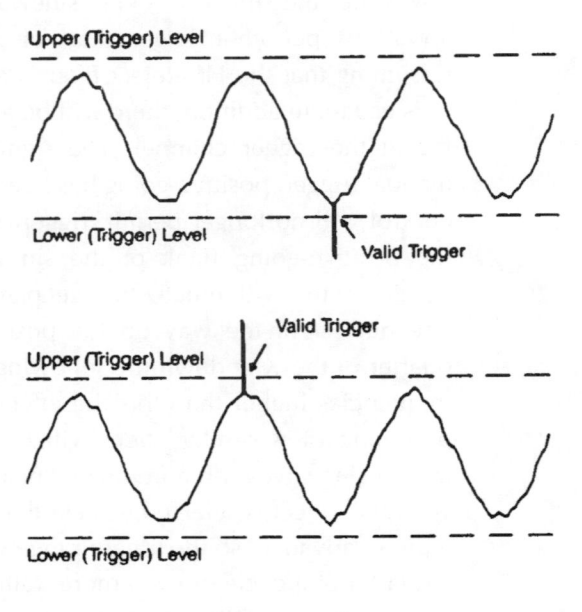

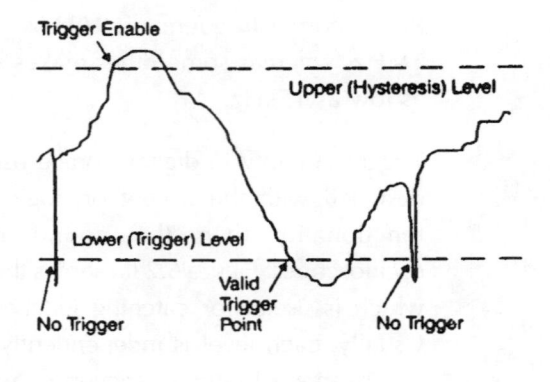

Digital storage oscilloscopes

Figure 6.2
In addition to a straightforward choice of level and polarity (perhaps with h.f. or l.f. reject), some oscilloscopes offer a variety of other triggering modes. Two are illustrated here. Window triggering (a) is useful in a DSO in 'babysitting' mode, waiting to capture an elusive glitch, while hysteresis triggering (b) can help with triggering on a noisy signal. (Reproduced by courtesy of LeCroy Ltd.)

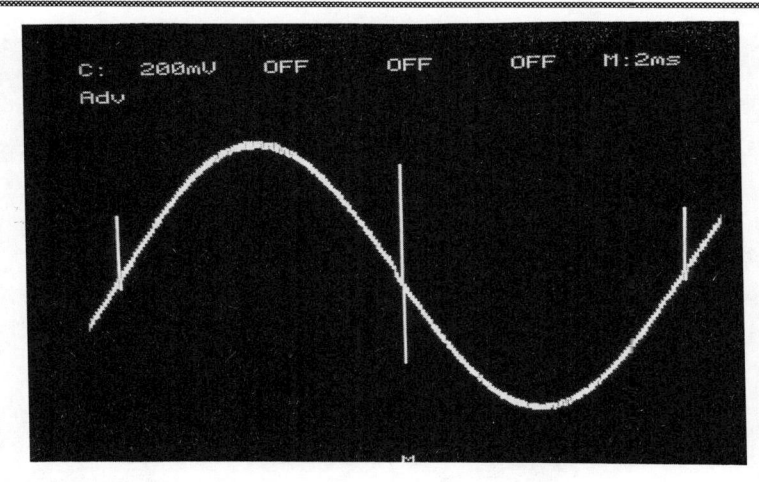

Triggering

Figure 6.3
True dv/dt glitch capture is a feature of some Gould digital storage oscilloscopes. Useful where a glitch may occur at any level on a waveform, without exceeding the level of either peak. (Reproduced by courtesy of Gould Instrument Systems.)

The user can define a voltage change between two points that is greater than the changes normally found in a sinewave. The arrangement is particularly useful for glitches on AC power lines, where the glitch may occur at any point on the waveform, but its amplitude will often not exceed either the positive- or negative-peak. In investigating three-phase applications, the multichannel function will trigger the DSO if there is a glitch on any of the phases, see Figure 6.3. On some other makes and models, the same effect can often be achieved by using the *trigger on pulse width less than* (the user set amount) facility, provided the glitch crosses the set trigger level. Clearly, the smaller the glitch amplitude the less likely this approach is to be successful, unlike a true dV/dt trigger mode.

Conventional glitch triggering monitors the signal for pulses less than a specified width. There are several possibilities, depending on the particular instrument. It may monitor for pulses that would fall between successive sample points at the selected timebase speed. This is achieved by running the ADC at its maximum rate, even though only every hundredth or thousandth sample may be stored at the selected timebase rate. Alternate storage locations may be dedicated to the highest and the lowest sample values encountered in the sample period, giving a MAX/MIN or ENVELOPE mode of operation; this is very effective at catching

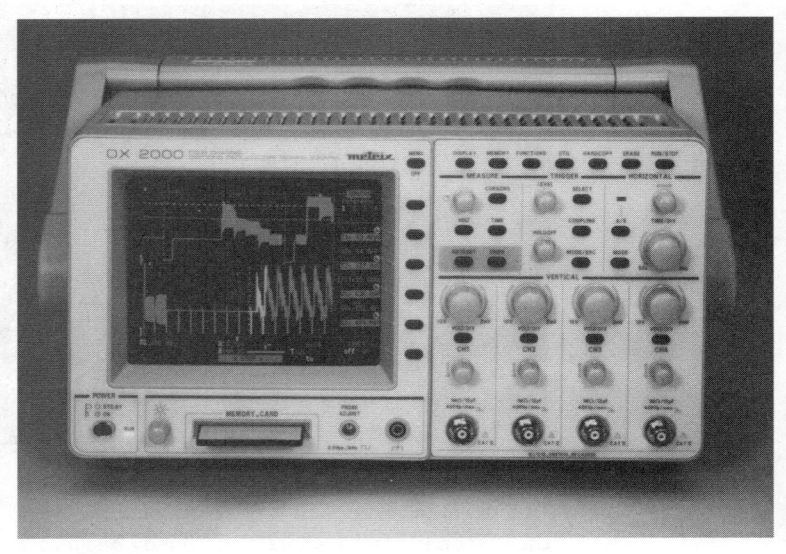

Digital storage oscilloscopes

Figure 6.4
The METRIX model OX2000 is a four channel digitising oscilloscope with a 150MHz bandwidth, providing 200Ms/s on Channel 1 and 100Ms/s on the other three channels. Acquisition modes include Roll, Direct, Envelope, Max/min and glitch capture. (Reproduced by courtesy of METRIX S.A.)

the fastest glitches that the instrument can see. If a glitch is so fast that it may fall between samples even at the ADC's maximum sampling rate, it may still be possible to capture it on some instruments. These are fitted with dedicated glitch capture circuits which register glitches down to a few nanoseconds wide, stretching them so as to permit them to be digitised by the ADC. On the other hand, interval triggering monitors the signal for pulses wider than a user specified width. It is useful for capturing signal drop-outs or undersized *runt* pulses. Many digital storage oscilloscopes offer a facility whereby the trigger is delayed either by a user specified time, or by a user specified number of trigger events. This permits the user to view in detail specific waveform sections without requiring extreme lengths of waveform memory. It is especially useful in conjunction with pattern triggering for testing digital systems. Pattern triggering lets the user select levels and slopes for several inputs. A trigger occurs only when all conditions are simultaneously met. Triggering can be set to occur when all the set conditions first occur simultaneously, or when thereafter any one of them ceases to meet the set conditions.

Chapter 7

Signal processing

The signal processing available in digital storage oscilloscopes takes one of two basic forms; pre-processing and post processing. Pre-processing takes place at acquisition time while post processing is performed upon the stored data in the display memory. Another distinction is between processing which is applied on a per channel basis, and processing that inherently involves two channels. Examples of processes applied on a per channel (per trace) basis are invert, average, envelope (max/min), smooth, interpolate, rms, integrate, differentiate, square root, log, absolute value and exponentiate. Processes involving two channels include add, subtract (add with one channel inverted), multiply and divide. Where one trace represents the voltage across a circuit and the other the current through it — e.g. by means of a current probe — then the multiply function will provide a third trace showing the instantaneous power in the circuit; additional function (waveform processing) memories are commonly provided for the results of processing on single or two traces. For the product of voltage and current traces to be a true representation of the power in a circuit, the two channel inputs should be dc coupled (unless it is required to measure only the varying component of the power) and any input variable offset control should be set to the OFF (zero voltage) position.

Averaging and smoothing are important techniques, as they can both reduce the effect of noise accompanying the wanted signal, and hence improve the signal to noise ratio. Averaging is used for reducing noise on multiple acquisitions. Smoothing can also be used with repetitive acquisitions, but it acts on each acquired waveform independently. It can, therefore, unlike averaging, be used on a single shot acquisition. Smoothing is a filtering algorithm which averages the values of (typically) five consecutive points on the waveform, and leaves the result at the centre point. It then moves on one point and repeats the process. Note that the five points averaged at each successive application of the algorithm are not all 'raw' sample values. The two earliest points are themselves smoothed values and so were in their turn averaged with even earlier points, as shown in Figure 7.1. By 'smearing' five points together, smoothing is very effective at reducing random high frequency noise on the display, but as Figure 7.1 shows, it costs you bandwidth. The reduction in bandwidth can be largely prevented from affecting the wanted waveform by sufficiently increasing the number of sample points per cycle of the waveform, viz by selecting a faster timebase speed, so that fewer cycles are displayed across the screen.

Averaging differs from smoothing in that the points averaged all occur at the same point of the waveform at each repetition. Therefore there is no reduction in bandwidth and the improvement in signal to noise ratio is independent of the frequency of the noise (within certain limits). During any triggered acquisition, a particular sample's amplitude has two parts: a signal component and a noise component. Because the incoming signal has a fixed relation to the trigger point, the signal component's amplitude remains the same from one repetitive waveform acquisition to the next. Random noise, on the other hand, has no fixed time relationship with the trigger point. The noise contribution to a particular sample's amplitude may be positive on one acquisition and negative on another, with an average of zero in the long run. Thus the greater the number of acquisitions averaged, the greater the noise reduction.

The foregoing explanation of the signal to noise ratio improvement, offered by averaging, sounds convincing at first

Signcl processing

smoothed waveform (partial)

section of waveform record, acquired single shot

A B C D E F

one smoothing set yields: $C' = \dfrac{A' + B' + C + D + E}{5}$

the next smoothing set yields: $D' = \dfrac{B' + C' + D + E + F}{5}$

(a)

smoothing algorithm

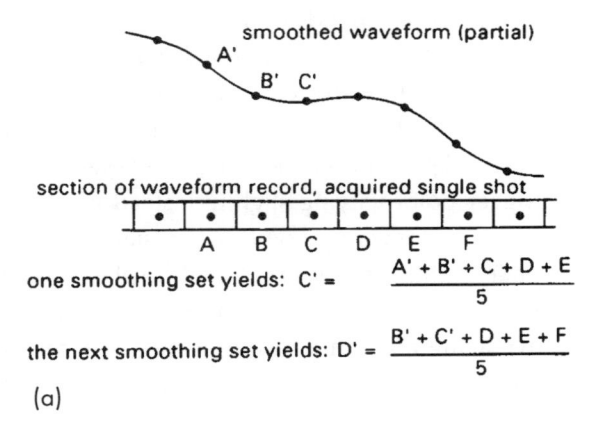

$$Y_n = \frac{1}{N}\left(\sum_{i=1}^{M} Y_{n-i} + \sum_{i=0}^{M} X_{n+i}\right)$$

where

N is the set of points processed; in this case, N = 5 (two points on either side of the centre point).
M is the number of points used on each side of the centre point, that is (N−1)1/2; in this case M = 2.
n is the horizontal position, represented by a point in the record (0 to 1023).
Y_n is the vertical value (result); the output array
X_n is the acquired value; the input array.

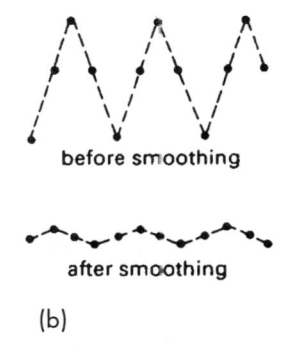

before smoothing

after smoothing

(b)

Figure 7.1
Waveform processing. In (a) smoothing moves through a waveform record, point by point. It averages each point with the two points behind and the two points ahead. It then leaves the average result at the centre point. As this figure shows, the first two values in each five-point average come from previously processed points. Smoothing under the worst-case sample-rate conditions shown in (b) reduces the triangle wave to a nearly straight line. (Reproduced by courtesy of Tektronix Inc.)

sight. But it depends upon one crucial assumption, namely that the signal component has a fixed relation to the trigger point. However, the trigger circuit is, unfortunately, usually not operated solely by the signal, but by the actual input, consisting of signal plus noise. The noise component will result in the actual trigger point varying slightly instead of always occurring at the intended triggering point on the signal. Thus the waveform will jitter slightly in the horizontal display direction. Thus, at each sample point, in addition to averaging out the noise, the averaging algorithm has to average out the variation caused by looking at a slightly varying point along the waveform. So advantage should be taken of any available means of cleaning up the signal applied to the trigger circuit, such as selecting the HF REJECT trigger coupling mode if appropriate, probably in conjunction with one of the advanced triggering modes available on many DSOs, such as hysteresis triggering etc., see Chapter 6. The waveform under investigation (with its attendant noise) may be the response from a circuit under test, the stimulus being a clean, noise-free signal. Clearly in this case, EXTernal triggering should be used.

The number of acquisitions averaged is usually user selectable in the range 2 to 256, as is the case of the digital storage adaptor (a digital storage oscilloscope less display, designed to work with a user's existing oscilloscope) illustrated in Figure 7.2. In this instrument, the user has the choice of two display modes. In one, the display is only updated when the selected number of traces has been averaged. As this could involve quite a wait if the selected number of averaged acquisitions is large, an alternative mode updates the display after each acquisition, to show the current resultant average. In either mode, when the selected number of acquisitions has been averaged, the instrument embarks on a new set with the 'slate wiped clean', to ensure that if the signal amplitude, waveshape or frequency changes, the current value will replace the previous completely. In the continuous update mode, changing the timebase frequency will reset the averages-taken counter, forcing the instrument to commence averaging a new set of traces. However, changing the Y sensitivity setting does not initiate a new averaging cycle, on the grounds that the actual input amplitude could change at any time, anyway.

Signal processing

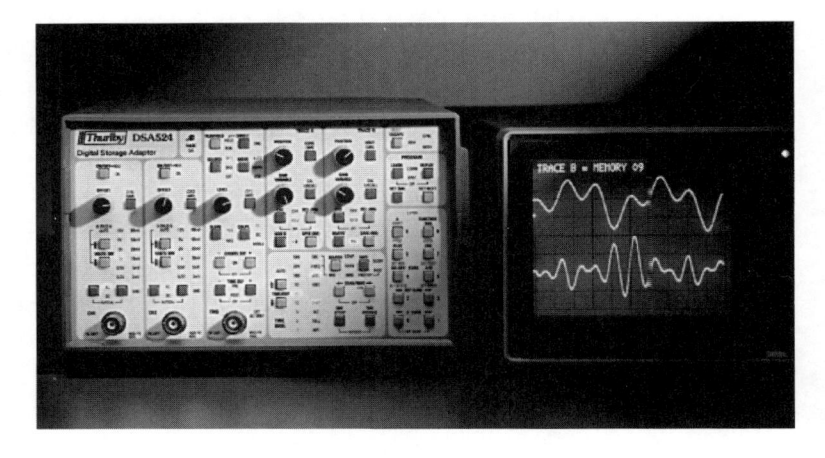

Figure 7.2
The Thurlby-Thandar DSA524 Digital Storage Adaptor — see text.
(Reproduced by courtesy of Thurlby Thandar Ltd.)

Other instruments use differing averaging algorithms, prominent among which are *exponential averaging* and *stable averaging*. These are both running averages over a user selectable number *n* of samples, but work rather differently. Consider exponential averaging, where the number of samples averaged *n* has been set at 16. Then the value of each sample actually displayed is calculated as 1/16th of the sample just taken plus 15/16ths of the corresponding sample displayed (not taken) on the previous acquisition. Thus any aberration from the true value of the signal due to noise on the sample will be reduced by a factor of 16. The currently displayed value is mainly determined by the last *n* samples, where n equals 16 in the sample just quoted. Because, with the arrival of a new sample, the value of the current sample is reduced to 15/16ths before adding 1/16th of the new sample, after 16 new samples it will be reduced to $(15/16)^{15}$ or 38 per cent. Now this is (approximately) e^{-1}, i.e. the effect of samples fades out exponentially with time over *n* samples, hence the name exponential averaging.

When the timebase speed or Y sensitivity setting is changed, with exponential averaging there would be a delay before the correct waveform is displayed, and this could be considerable if the

53

timebase speed is slow and *n* is large. To avoid this, some DSOs switch temporarily to stable averaging whenever one of the instrument settings is changed. Following the change, the first sample at each point is displayed at full amplitude, and will therefore show no signal to noise improvement. The second sample at the point is averaged with the preceding one on a 50/50 basis, one third of the third sample is added to two thirds of the second and so on until after *n* samples the algorithm switches back to exponential averaging. Thus stable averaging weights succeeding samples less and less heavily, i.e. the earlier samples have most effect, unlike the constant weighting of exponential averaging, and so the display rapidly converges on the new picture. The effectiveness of averaging in reducing noise is clearly illustrated in Figure 7.3.

Averaging not only decreases noise, it can actually increase the resolution of a DSO, unlike smoothing. Stable averaging increases the digitiser's potential resolution a factor of *n*, or ($\log_2 n$) bits, when *n* bits are averaged, although the actual increase in resolution is rather less than this. Exponential averaging provides the same improvement, but only after rather more acquisitions. Paradoxically, it is only the presence of the very noise on the signal which averaging is used to reduce, that provides the increased resolution, as a moment's reflection will show. For in an ideal noise free system, any given voltage within the input range of the ADC will always be digitised as the same value. Thus for an ideal 8 bit ADC, a constant mid-range voltage that digitises as 128 will always digitise as 128, even though its actual value is anywhere between voltages corresponding to 127.5 and 128.5. Now imagine however that there is just one digit peak to peak of random noise riding on the signal. An input voltage which ideally should digitise as 128 will still do so on average. However, an input which should ideally digitise as 127.5 will now digitise as 127 as often as 128. So if we add sixteen samples and divide the answer by 16, the odds are (statistically) that we will get 127.5, which can be represented exactly as a nine bit result. Furthermore, a level which should, in an ideal system, digitise as 127.75 will on average digitise as 128 three time as often as 127. This can be represented exactly as a 10 bit result.

Due to the statistical nature of noise, the signal to noise ratio (SNR) improves with increasing *n* rather more slowly than the

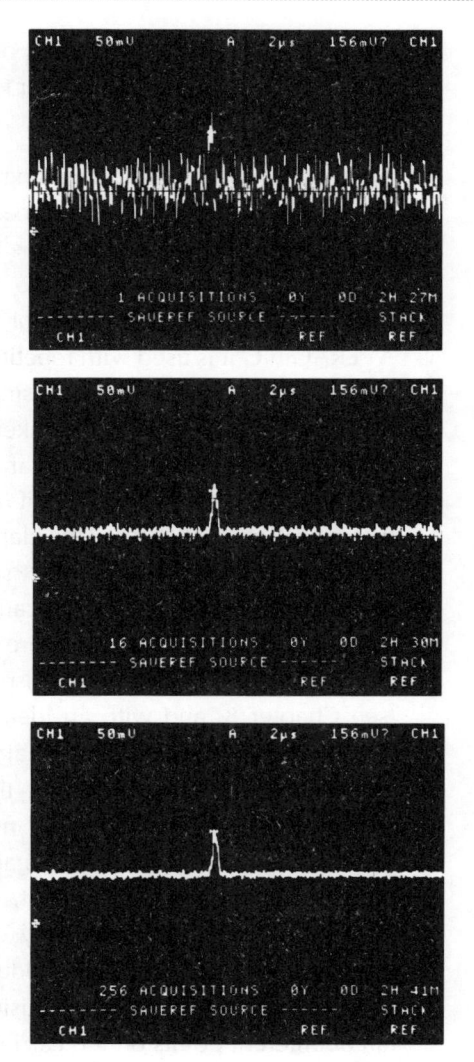

Signal processing

Figure 7.3
These photographs show how averaging cleans up the display of a spike that is nearly completely obscured by noise. (Reproduced by courtesy of Tektronix Inc.)

potential increase in resolution. In practice, the SNR improvement is just 3dB per doubling of the number of averaged samples n, i.e. the effective number of extra bits is $\frac{1}{2}\log_2 n$, or 4 bits for 256 samples. Bearing in mind the requirement for 1 bit peak to peak of noise (1 bit loss of accuracy) to make it all happen, 256 samples can improve the accuracy of an 8 bit system to $(8-1+4) = 11$ bits. With less than one bit of noise, the improvement will not be

Digital storage oscilloscopes

obtained, while with more than one bit of noise, more than 256 samples will be required in order to drive the higher level of noise down as far.

Another type of signal processing uses two storage locations for each point displayed on the screen, or on many models of DSO, for economy, uses conjointly the two storage locations for each pair of adjacent points for the same purpose. This is called variously MAX/MIN MODE, or ENVELOPE MODE and like AVERAGING it is used with repetitive waveforms, although unlike that mode it can be used for a single shot acquisition. When the first sample of a new sweep is taken, it is compared with the value already stored in that location, and discarded if smaller but used to overwrite the previous value if larger. Thus, over many sweeps, that location will store the largest value (the MAX value) encountered. Similarly, the next sample is used to overwrite the value in the next storage location only if it is smaller (the corresponding input voltage more negative), thus storing the MIN value. The stored waveform is displayed in DOT JOINING MODE (see Chapter 8) and with an ideal noise free waveform with no frequency components above half the sampling rate, the display will look the same as without the ENVELOPE MODE in use. However, if the waveform is noisy or unstable, or there is significant energy at frequencies above the Nyquist rate, then the picture can look quite different, see Figure 7.4. In the middle photograph, the high frequency carrier is incorrectly shown as aliased to only 20 times the modulating frequency, in this single shot picture. In successive acquisitions, the sample points would fall at different points on the carrier (assuming it was not a locked harmonic of the modulating frequency), giving constantly changing pictures, each very similar to that shown.

In the envelope mode display, lower photograph, the maximum and minimum values at alternate positions in the X direction are joined by vectors, giving a display very similar to the non-storage display on an analog oscilloscope, top photograph. With the usual density of 100 dots per horizontal graticule division, the result is a band of light indicating the envelope of the maximum and minimum values encountered. The envelope mode leads conveniently into one of the most important features of the digital storage oscilloscope: *glitch capture.*

Signal processing

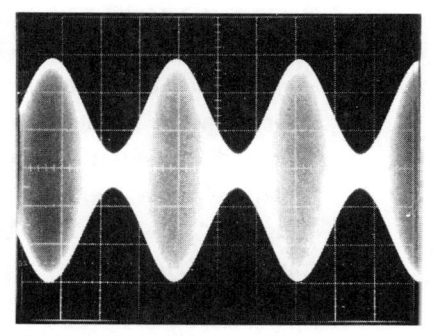

NON-STORAGE

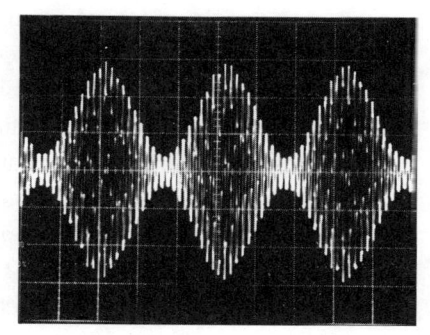

NORMAL MODE

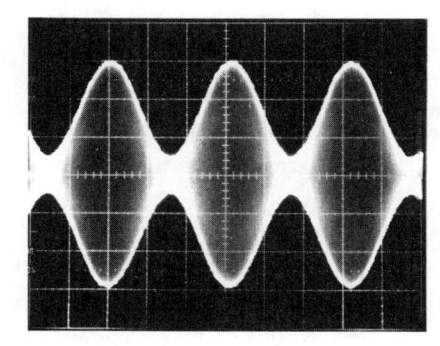

ENVELOPE MODE

Figure 7.4
These photos are of an amplitude modulated signal as it was displayed by a non-storage scope, by a digital storage scope in normal mode and by a digital storage scope using the envelope mode. The modulating frequency is reproduced easily in both digital acquisitions. The carrier, however, is being digitized at a rate much less than two samples per period and is shown as a lower frequency in the middle photograph. The envelope mode used as an anti-aliasing feature results in a display very much like the non-storage signal. (Reproduced by courtesy of Tektronix Inc.)

Digital storage oscilloscopes

A *glitch* is a rogue narrow pulse which can play havoc in digital systems. It is typically due to a rare condition and depending on the previous pulse sequence, may only appear occasionally — often with dire results — making it very difficult to observe. If the triggered waveform is acquired repetitively for a long period in envelope mode, then with luck the glitch, when it occurs, will be caught as an isolated positive level sample standing up from the logic 0 level (0 volts), or as a negative-going glitch from the 1 level (+5V or +3V as the case may be). However, if the time per division setting is, say, 10ms/div, then with 100 points per division, a glitch of width 100ns or less could very easily be missed. A powerful enhancement of ENVELOPE MODE for digital glitch capture is to run the ADC sampler at its maximum rate, regardless of the actual time per division setting. Thus in the example just given, the sampling rate would be not just 10Ms/s (100 samples per 10ms), but, say, 50, 100 or more Ms/s, whatever the DSO's maximum sample rate is. However, only one sample value (the highest or lowest in each alternate MAX or MIN storage location) is recorded in each 100ns sample period. By running the digitiser at its maximum rate in ENVELOPE MODE, regardless of the timebase time/division setting, the optimum digital glitch capture performance is always obtained. The DSO can therefore be left 'babysitting', just waiting for a glitch to occur, with the certainty that it will be captured — provided that it is not so narrow that it can slip between samples at the instrument's maximum digitising rate.

In high speed logic circuitry, notably ECL logic, glitches as narrow as one or two nanoseconds can occur. Even on one of the more expensive DSOs capable of digitising at 500Ms/s, the digital glitch capture mode described above could not guarantee to capture that, let alone an oscilloscope with a 100Ms/s maximum digitising rate. However, there is another approach, using analogue peak detectors, which enables lower priced instruments to capture otherwise elusive glitches. These detectors, in effect one for each polarity, stretch any narrow pulse, holding its maximum value until digitised, after which the detector circuit is reset ready for any following glitch. They are thus an example not just of pre-storage processing, but of pre-digitising processing. This scheme is used in many instruments which are still in service,

Signal processing

but the need for it is being overtaken in the latest models, where blindingly fast digitising rates of 1Gs/s and above make it redundant.

Figure 7.5
The Gould model 840 four channel high resolution digital storage oscilloscope features a 1MHz bandwidth in 12 bit resolution mode with, unusually, an anti-alias filter selected in accordance with the timebase setting. In 8 bit mode the maximum single shot sample rate is 100Ms/s, or equivalently 2.5Gs/s in repetitive mode, giving a 150MHz bandwidth. In addition to internal RAMDISK, there are floppy or hard disk or memory card options. (Reproduced by couresy of Gould Instrument Systems.)

Dual mode oscilloscopes combine digital storage operation with the ability to display waveforms without digitising them first, they are thus in effect two oscilloscopes in one: a DSO and a conventional analog oscilloscope. This implies that they use an electrostatically deflected high performance oscilloscope tube, with all the expense that involves. In DSO mode, such dual mode oscilloscopes usually use vector scan display, which is described in more detail in Chapter 8. However, they are definitely in the minority, most DSOs having no such direct analog capability. This means that they can use a magnetically deflected cathode ray tube, either monochrome or colour, of the type used in small TV sets or monitors, and of course raster scanned.

A mode of operation often found in such DSOs, especially those at the upper end of the market such as the Hewlett-Packard 54000 series, is infinite persistence. In a digital storage oscilloscope without this facility, it is only necessary to store, for each of the (typically) 1024 points across the screen, a digital word corresponding to the displayed waveform's voltage at that point, for the displayed trace of each channel. In infinite persistence mode, much more display memory is needed, although the acquisition memory requirement is unaffected. Instead of the waveform data on a scan replacing that gathered on the previous scan (or being averaged with it), it is displayed in addition to the previous trace. With an ideal noise free repetitive signal, just a single trace will appear, but if the signal is noisy, the trace will appear widened or blurred, just as on an analogue oscilloscope.

This implies that the display memory must be able to store (at least) one bit of information for every possible vertical point on the display — a bit-mapped or pixel oriented display is needed. A one-bit bit-mapped display will only permit a monochrome display of the signal, each point of the display being either ON or OFF, depending on whether that particular voltage level has been recorded at that particular point across the screen. If the display memory can store a byte of data for each pixel, then the intensity of each point of the display can be controlled to give a 256 grey level rendering, indicating how many times over the last 256 scans that particular value of voltage was recorded at that particular point across the screen, i.e. at that particular point on the waveform. Alternatively, where a colour display tube is used, the intensity of the trace may be maintained constant. The frequency of occurrence of any voltage level at a particular point on the waveform is indicated by the hue — e.g. from red for the most frequently occurring values, through the spectrum to blue for the least, and of course black for voltage levels which have not occurred at all. Figure 7.5 illustrates infinite persistence mode in use, showing intermittent partial transitions and double transitions at the output of a D flip-flop. Another use for this mode of operation is displaying eye diagrams.

Eye diagrams are a powerful means of evaluating the distortion suffered by digital data in passing through a transmission path. Originally a nice tidy data stream of noughts and ones with fast square transitions, after passing through a band limited channel

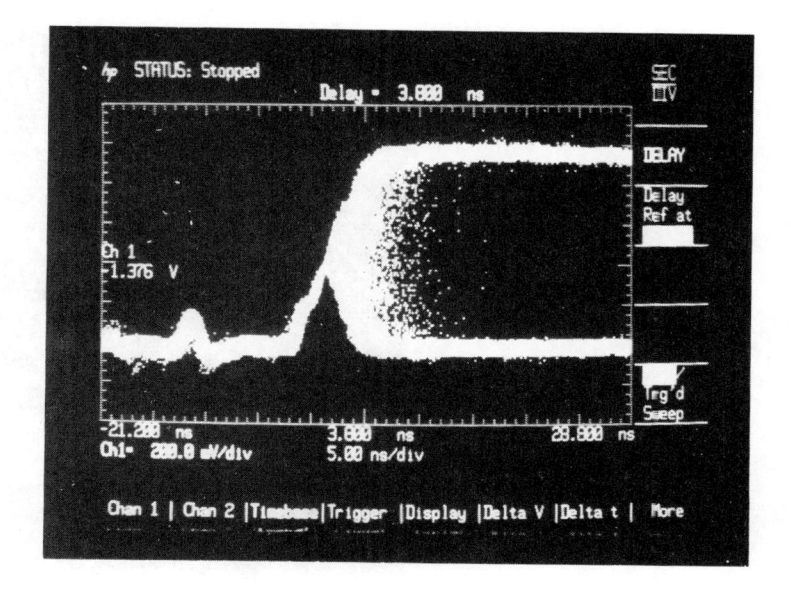

Signal processing

Figure 7.6
The HP54000 series DSOs use a full colour, graphics-type display with pixel storage. This makes it possible to store and display all voltage levels occurring at each point along a waveform over many repetitions. This is called the *infinite persistence* mode and the illustration shows its use to display intermittent partial transitions and double transitions at the output of a D flip-flop. (Reproduced by courtesy of Hewlett-Packard Ltd.)

(bandwidth is a scarce and hence expensive commodity), the data emerges looking almost sinusoidal. It should however be capable of furnishing the original data stream when sliced by a comparator and then retimed by the recovered data rate clock, which is extracted from the incoming data itself. If the noise is excessive (or the transmission system poorly designed), the eye will appear partially or largely closed, and the corresponding recovered data will exhibit a high BER (bit error rate), this is illustrated in Figure 7.7.

An eye diagram consists of a display of the incoming raw data, bandlimited and corrupted by noise as it is, the display being triggered by the recovered clock. This is shown diagrammatically in Figure 7.8 (a). With the digital signal processing (DSP) power built into modern DSOs, it is possible to derive more information than ever from an eye diagram and in particular to make quantitative measurements of the signal's statistical properties.

Digital storage oscilloscopes

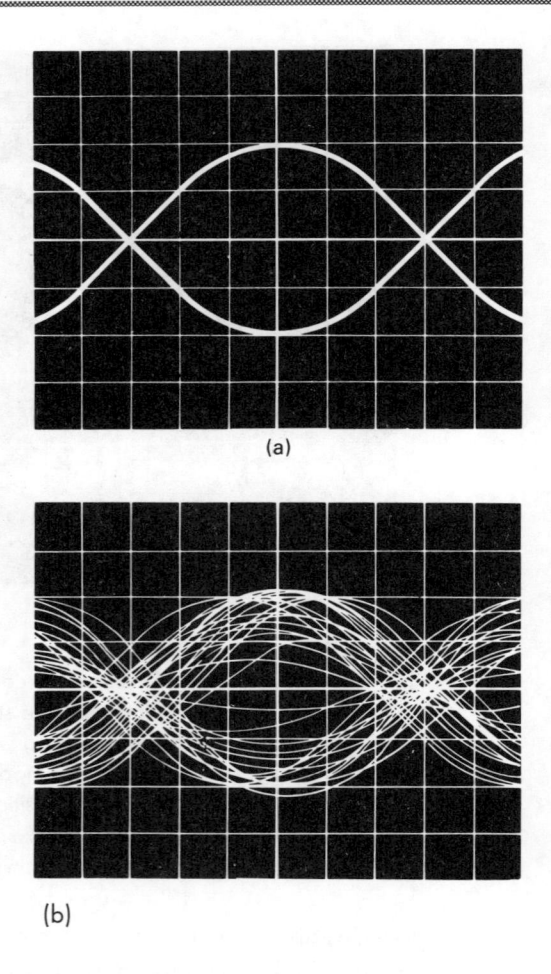

Figure 7.7
Two-level digital phase modulated signal showing (a) well-setup system with no intersymbol interference (b) poor system with bad intersymbol interference

This permits correlation between the measured degradation of the raw signal and the BER of the recovered data, a thing which was previously extremely difficult if not impossible. Figure 7.8 (b) shows how with a bit-mapped display with 16 bits per pixel, the instrument can over a period, totalise the number of sample points falling in each pixel. The resultant eye diagram is displayed in colour with, say, single or low count pixels shown in shades of blue, through the spectrum to red for pixels with the highest counts. Additionally, the data can be further processed in various

Signal processing

ways to show histograms illustrating the 'openness' of the eye, see Figure 7.8 (c).

Finally in this chapter a word about sine interpolation display techniques. This topic is covered here rather than in the next chapter, which deals with display techniques among other things, because sine interpolation involves a fair amount of signal processing to prepare the data for display.

The maximum bandwidth of a DSO is set ultimately by the bandwidth of the input circuitry, including the Y input amplifier. Given a fast enough sample and hold preceding the ADC, signals up to this frequency can be captured using *equivalent time sampling*, also known as *repetitive mode*, assuming always that the waveform is repetitive. But if the signal of interest is a unique event, this option is not available, and the DSO must be operated in single shot mode. Bandwidth is now limited by the maximum sample rate of the ADC, since about ten samples per cycle is normally considered the minimum number for satisfactory rendition of the waveform, as illustrated in the next chapter. Sine interpolation can help here, subject to certain conditions.

A theorem due to Nyquist states that to define a sinewave, a sampling system must take more than two samples per cycle. It is often stated that two sample per cycle are necessary, but this is not quite correct. Exactly two samples per cycle (usually known as the *Nyquist rate*) suffice if you happen to know that they coincide with the peaks of the waveform, but not otherwise, since then although you will know the frequency of the sinewave, you have no knowledge of its amplitude. And if the samples happen to occur at the zero crossings of the waveform, you would not even know it is there. However, with more than two samples per cycle — in principle 2.1 samples would be fine — the position of the samples relative to the sinewave will gradually drift through all possible phases, so that the peak amplitude will be accurately defined.

A good sine interpolator can manage with as few as 2.5 samples per cycle, always assuming of course that the waveform being acquired is indeed a sinewave — but if it were, it would usually be repetitive, so that single shot acquisition was not necessary.

Digital storage oscilloscopes

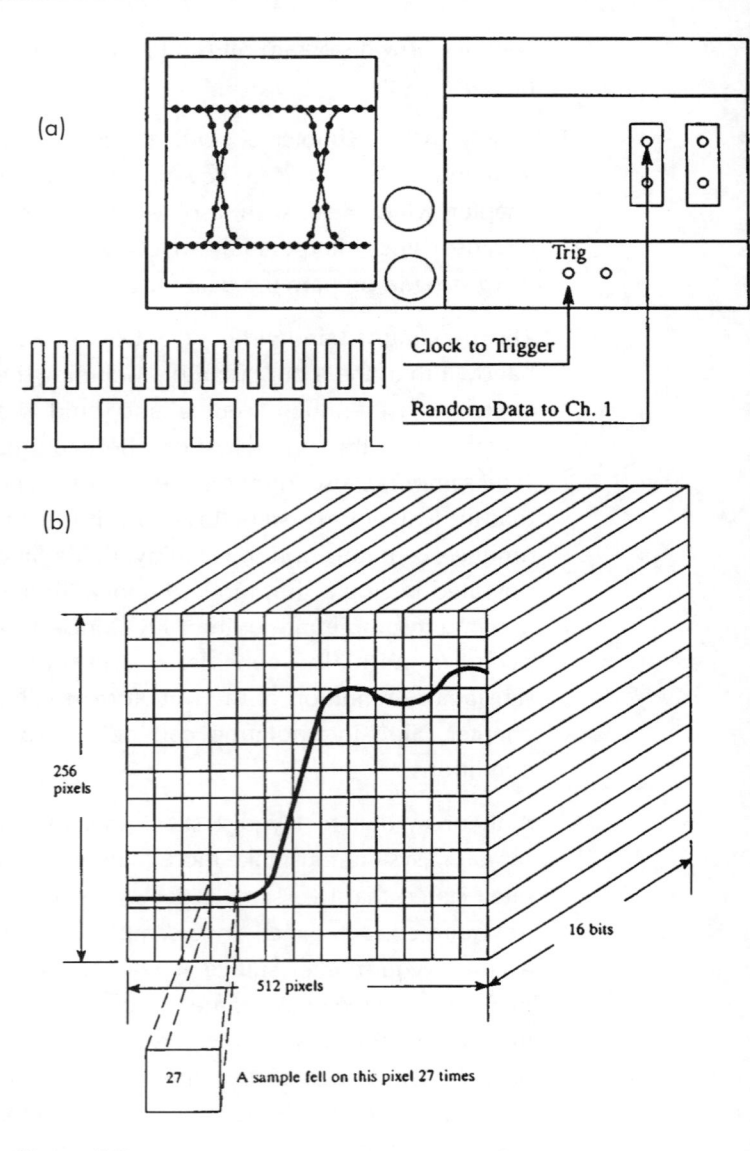

Figure 7.8
Measurements on eye diagrams, see text. (Reproduced by courtesy of Tektronix UK Ltd.)

The exception is an isolated burst of several cycles of a sinewave. For non-sinusoidal waveforms a sine interpolator is usually inappropriate, except in the case of certain instruments which can suitably pre-process the waveform before passing it to the sine interpolator. The effect of a sine interpolator on a step waveform

Signal processing

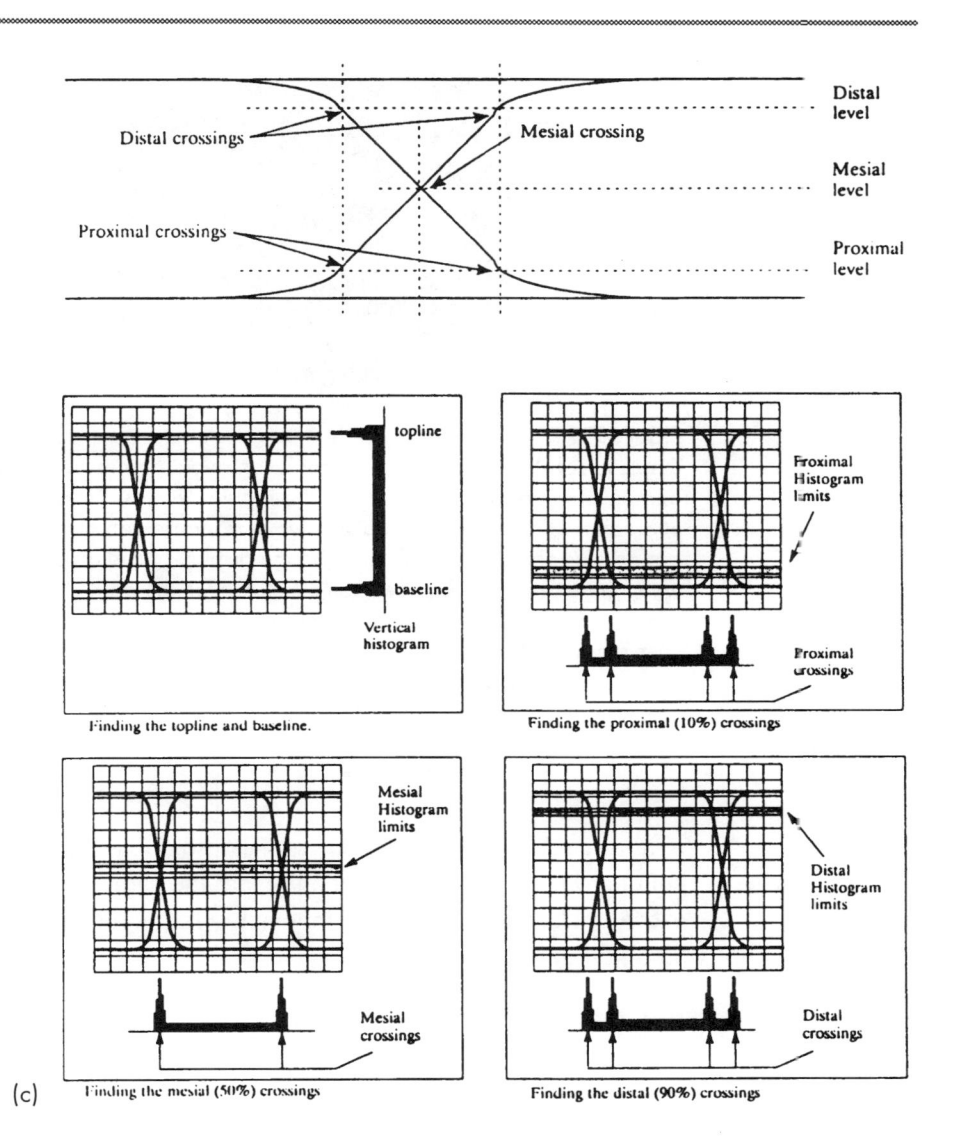

(c)

Figure 7.8 (contd)

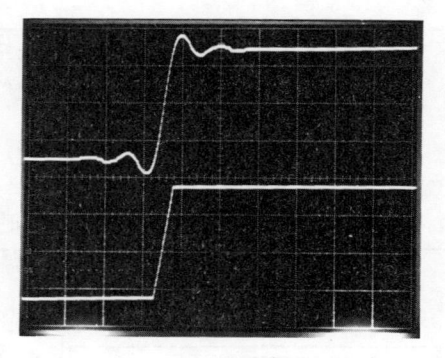

Digital storage oscilloscopes

Figure 7.9
Displays constructed with sine interpolation avoid perceptual aliasing and envelope errors when used to display sine waves. But an interpolator designed for good sine wave response can add what appears as pre- and over-shooting to the display of a step function when there are less than three samples taken on the step. The error is minimized if more than three samples are taken and with narrow spectrum waveforms such as sine waves. The photograph is a double exposure of a signal with no samples on the step: the first trace is drawn with a sine interpolator and the second with a pulse interpolator. (Reproduced by courtesy of Tektronix Inc.)

is illustrated in Figure 7.9. For non-sinusoidal waves, accurate definition of the waveform requires that the sample rate should exceed twice the frequency of the highest harmonic of significant amplitude.

Chapter 8

Displays and display technology

The different types of display used in digital storage oscilloscopes are closely interlinked with the type of display technology used. The main types of display devices are LCD (liquid crystal device) panels and cathode ray tubes. Also occasionally found are electroluminescent panels, though these have not proved popular and are currently only available in monochrome types. Both LCD displays and cathode ray tubes used in digital storage oscilloscopes are generally raster scanned, thus building up the displayed picture in exactly the same manner as a TV set. LCDs and cathode ray tubes are found in digital storage oscilloscopes in both monochrome or colour types, though the latter have lower resolution due to the need to pack in triads of red, green and blue pixels. Thus the colour cathode ray tubes used in digital storage oscilloscopes use the same shadow mask technology as used in colour TV sets, and the colour LCD displays the same technology as used in miniature colour TVs and in laptop computers.

Digital storage oscilloscopes with colour displays in the Tektronix range form a notable exception. These use a monochrome high resolution monitor cathode ray tube with a whitish phosphor, behind the Tektronix proprietary *colour shutter*. Thus a colour display without any of the resolution limitations of a shadow mask tube is possible. Unlike the landscape orientation of the cathode ray tube in a TV set, the high resolution monitor tubes are mounted in portrait fashion and scanned vertically, providing high resolution in the vertical direction. The resolution in the horizontal direction is set by the number of lines, which is commensurate with the number of samples displayed across the tube.

The cathode ray tubes which have been mentioned so far are all of the magnetically deflected variety, limiting them to low frequency operation. Scanning rates for such tubes are typically below 100kHz, the scanned raster display being typically rewritten on the screen around 100 times per second. This is more than adequate for displaying waveforms which have been acquired and stored in memory, but would be quite unsuitable for use in realtime analog oscilloscopes. The latter use electrostatically deflected tubes, the design and operation of which is described in detail in the companion volume *Oscilloscopes — How to Use Them — How They Work* (ISBN 0 7506 2282 2), from the same publisher. Depending upon their degree of sophistication (and hence cost), such tubes are capable of displaying frequencies up to 5GHz though the front-end electronics of an analog oscilloscope usually limits the top working frequency to around 1GHz at most.

More modestly priced electrostatically deflected cathode ray tubes function up to 100MHz or more and these are used in an important class of instruments which have been mentioned already. These are the dual mode oscilloscopes, which can operate either as realtime analog scopes or alternatively in digital storage mode, making use of the same cathode ray tube for both types of operation. Depending upon the model, the bandwidth in analog mode typically exceeds the single shot bandwidth in digital storage mode, but the bandwidth in equivalent-time digital storage mode (for repetitive signals only) may well equal (or even slightly exceed) that in analog mode. These instruments have the important advantage that the danger of being mislead by an aliased display is greatly reduced, as the user can always switch to realtime analog mode and be sure of what is actually happening. On the other hand, in digital storage mode, they provide a pre-trigger view capability not available on realtime only instruments. Nevertheless, despite these very real advantages, dual mode oscilloscopes are offered by only a fairly small number of manufacturers. A selection of such instruments is illustrated in these pages, Figure 8.1 being one example.

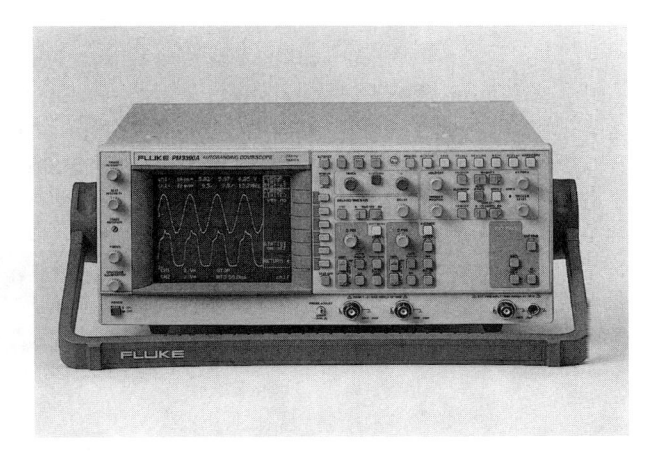

Figure 8.1
The PM3390A from Fluke is a dual-mode (realtime analog plus digital storage) oscilloscope. It offers a 200MHz bandwidth in analog operation, and an equal or greater bandwidth in the equivalent time mode of digital storage operation. The maximum sampling rate is 100Ms/s, giving a 20MHz single shot satorage bandwidth using sine interpolation. (Reproduced by courtesy of Fluke Corporation.)

Dual mode instruments such as those just mentioned usually use a *vector scanned* display for displaying the stored traces acquired in digital storage mode. This means that the trace is drawn on the screen in exactly the same way as in realtime mode. The output of a digital to analog converter turns the stored data into a time varying voltage. This is applied via the Y output ampl fier to the Y deflection plates, while a display timebase deflects the spot horizontally at a steady rate. Between successive writings of the trace (or the various traces when displaying more than one channel), readout data is written to the screen by applying appropriate deflection voltages to the X and Y deflection plates to cause the spot to trace out the various characters to be displayed. Because the whole display only needs to be rewritten fifty or more times per second, this type of display is also within the capabilities of a magnetically deflected cathode ray tube such as is used in digital mode only oscilloscopes. Nevertheless, pure digital storage oscilloscopes with cathode ray display tubes opt for a raster scanned display, since this is more convenient when no realtime display capability is involved.

The LCD displays used in digital storage oscilloscope vary widely in size and resolution. The small monochrome types, used in handheld digital multimeters with a waveform storage capability, may have a resolution of only about a hundred points in each axis. On the other hand, the colour LCD displays used in some top-of-the-range instruments may feature five hundred or more points in each axis.

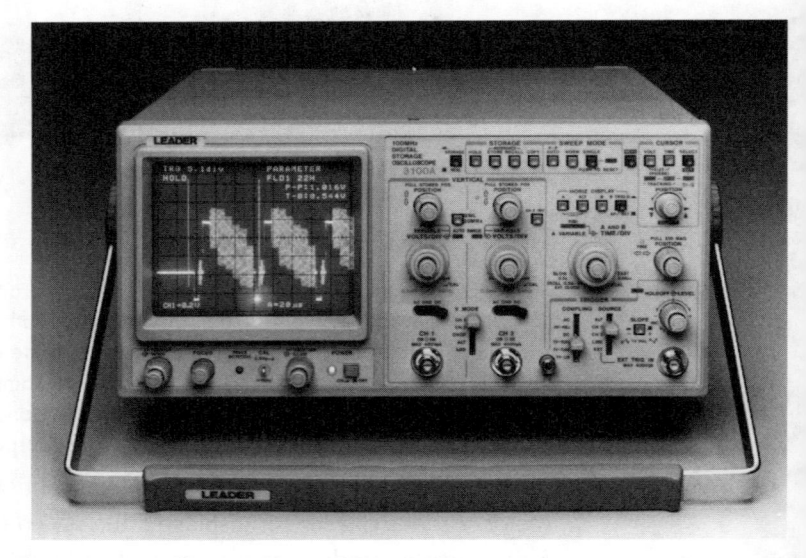

Figure 8.2
Another dual-mode oscilloscope, the two channel model 3100A from Leader provides 100MHz bandwidth in analog mode and also in digital storage mode, using repetitive sampling – bandwidth in single-shot mode is set by the maximum sample rate of 40Ms/s. The instrument features Dual A and B timebases for delayed measurements, TV triggering form Line, Field 1 or Field 2, cursor measurements and an exceptional sensitivity (at reduced bandwidth) of 500µV/division. (Reproduced by courtesy of Leader Instruments (Europe) Ltd.)

Having described the various display devices used in digital storage oscilloscopes, it is time to turn to the way the stored waveform data is presented on them. The simplest way is to use the output of the DAC (digital to analog converter) which is used

Displays and display technology

to reconstruct the stored data to define the vertical position of an illuminated point on the screen, the horizontal position being defined by the point along the waveform to which it refers. In a vector scanned display, the DAC output for each successive storage location would be applied (via a Y deflection amplifier) to the Y plates of the cathode ray tube, while the steadily incremented output of a timebase DAC would applied, via the X deflection amplifier to the X plates. This causes a series of points to be illuminated, delineating the stored waveform as a dotted line, as shown in Figure 8.3, top row, the trace being blanked between dots. (Note that the left hand columns show a display much expanded in the X direction; normally there would be many dots per horizontal division, as in the right hand column.) As can be seen, when there are too few dots per cycle, the display is difficult if not impossible to interpret. If the X deflection moves steadily and smoothly instead of being stepped, the dots are elongated into dashes, as illustrated in Figure 8.4, though in practice the vertical transitions between dashes shown would be deliberately blanked or in any case too feint to see.

With a raster scanned display, the same effect would be obtained by gating the output of the reconstruction DAC with the current line of the scan in a window comparator, the trace being blanked unless the comparator output is at logic 1 level. The same arrangement can be used to display the graticule, there being no graticule permanently marked on the face of the tube as in a realtime analog oscilloscope. Thus the displayed traces are always accurately located relative to the graticule, regardless of the horizontal and vertical linearity of the raster, an arrangement found, for example, in digital storage oscilloscopes from LeCroy and in some from Metrix (but not others). In a digital storage oscilloscope using a vector scanned display, the graticule may also be drawn electronically. But vector written displays are most commonly found in dual mode analog/digital storage oscilloscopes, and in these a permanently marked non-electronic graticule is common, for example in dual mode oscilloscopes from Fluke (including ex Philips models) and from Kikusui. In analog mode, the display accuracy of such oscilloscopes, measured against the graticule, depends upon the linearity of the

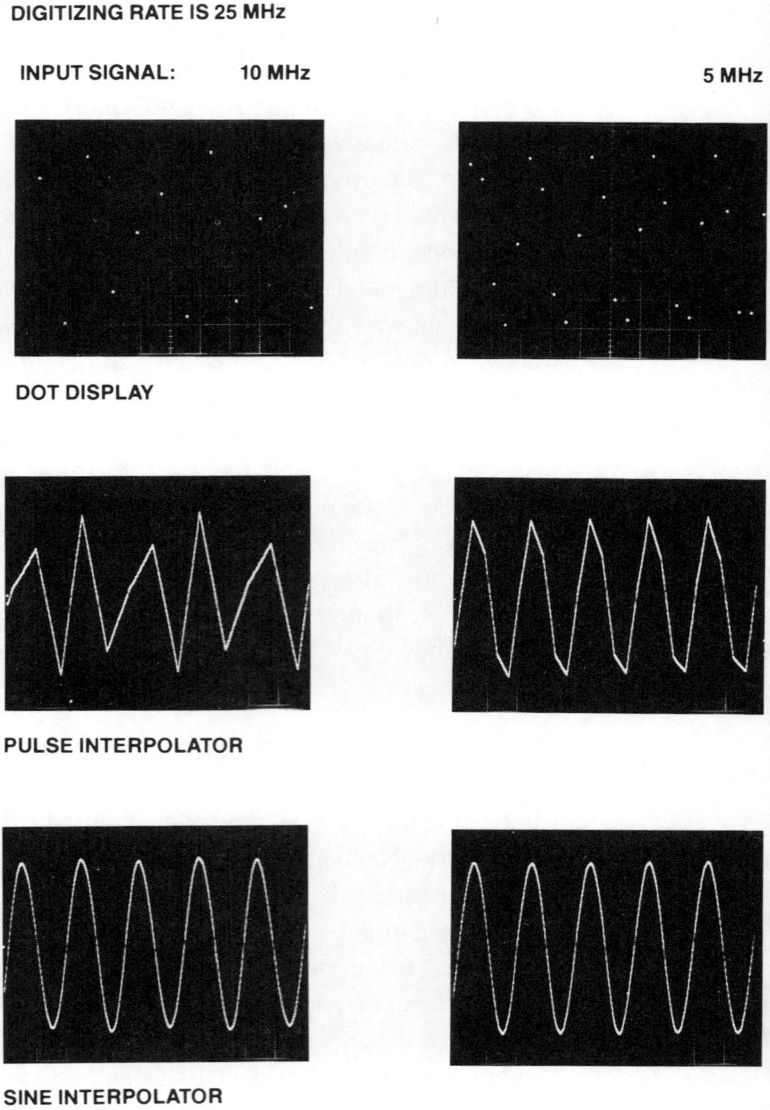

Digital storage oscilloscopes

DIGITIZING RATE IS 25 MHz

INPUT SIGNAL: 10 MHz — 5 MHz

DOT DISPLAY

PULSE INTERPOLATOR

SINE INTERPOLATOR

Figure 8.3
See opposite

X and Y amplifiers and of the cathode ray tube deflection geometry. The same comment applies to the display in digital storage mode, but where time and voltage measurements are made by means of the instrument's moveable cursors and their

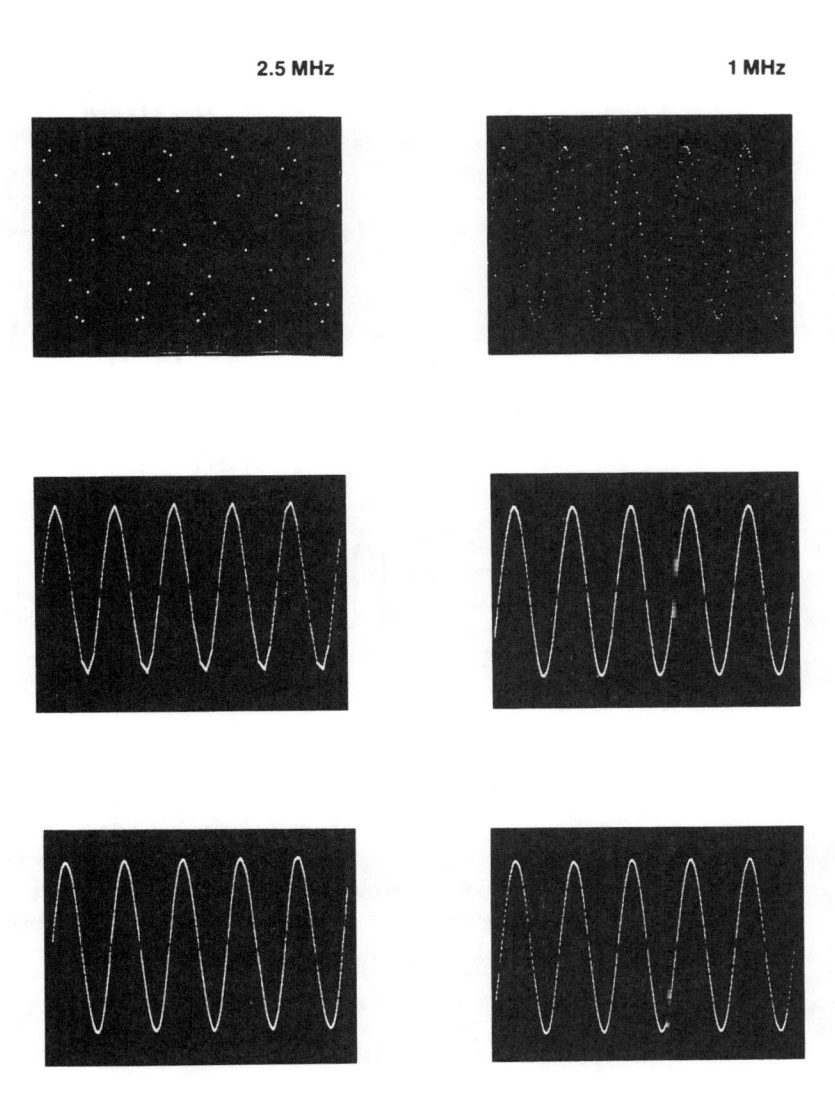

Displays and display technology

2.5 MHz **1 MHz**

Figure 8.3
The display reconstruction type influences the useful storage bandwidth of a digital scope. To trace a recognizable sine wave takes at least 20 and preferably 25 samples/cycle with dot displays. Pulse-interpolator displays produce a useful trace with about 10 vectors per cycle; peak errors make your measurements more difficult when fewer are used. The sine interpolator in the Tek 2430 display shown in the lower series of diagrams reproduces sine waves with only 2.5 sample/cycle, finally approaching the limits that the sampling theory suggests. (Reproduced by courtesy of Tektronix Inc.)

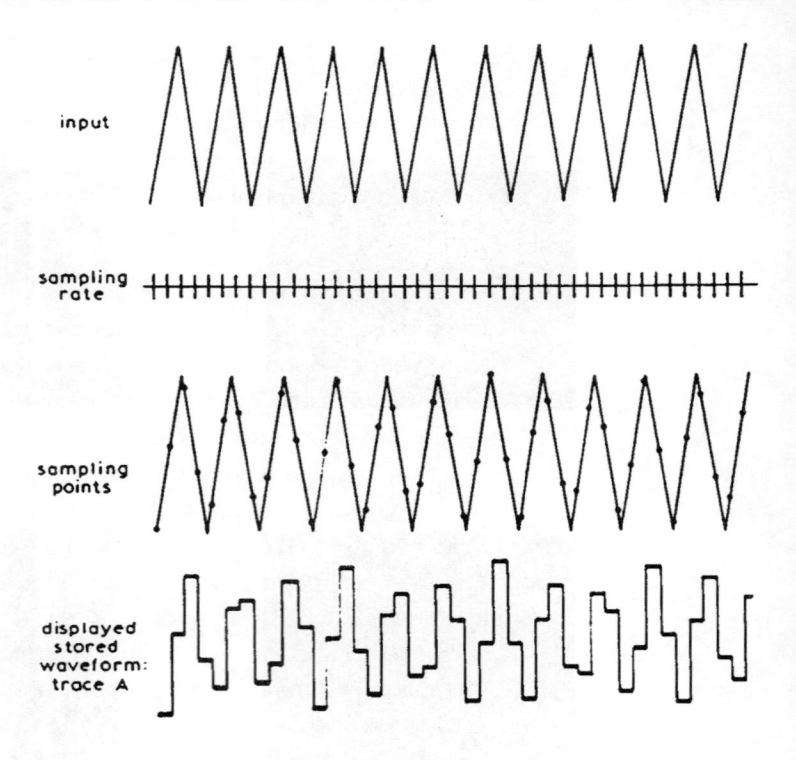

Digital storage oscilloscopes

Figure 8.4
Waveforms as displayed on a digital storage oscilloscope; the sampling rate is just less than four times the signal frequency. Trace A is the usual type of display: current sample held and displayed until replaced by next sample. The transitions between dots here shown as feint vertical lines would in practice be invisible. The dots themselves would be shorter than shown (with typically 1024 point horizontal resolution), except when using horizontal expansion.

associated readouts, these accuracy limitations do not apply.

Returning to the question of display formats, a distinct advance upon the dot display, when there are few samples per cycle of the waveform being viewed, is the technique known as dot joining, also called pulse interpolation, see Figure 8.3, middle row. Here, the display is much easier to interpret, although it only begins even to approach an accurate rendition of a sinewave when there are ten or more samples per cycle. Note also that the illustrations are of a single shot. The near-sinewave display shown at ten cycles per sample would be seen only if the 2.5MHz sinewave

Displays and display technology

were phase locked to the 25MHz sample rate. If the 2.5MHz sinewave frequency were 10Hz removed from one tenth of the sample rate, the waveform would appear to wobble, like a jelly on a plate, as the waveform drifted through all possible phases, while with a larger frequency offset, the waveform would simply appear blurred. This is illustrated in Figure 8.5, top photograph.

Clearly, for a dot joining display, the deflection in the vertical or Y direction must be smooth and continuous, rather than stepped, as it could be for a dot display, otherwise a stepped display as in Figure 8.C would result. The circuitry for changing the vertical or Y deflection smoothly from one dot position to the next is typically as in Figure 8.6. Two multiplying DACs are used, each with a sawtooth waveform (at the display dot sample rate) applied to its reference input. The sense of the two sawtooth waveforms is opposite, one decreasing as the other increases. The current sample value is latched and applied to the digital input of one DAC (the P DAC) and the next sample value to the Q DAC. The DAC outputs are summed, and as the contribution from the P DAC falls to zero while that from the Q DAC rises from zero to the next sample value, the sum changes linearly from the current value to the next. The next sample value is then transferred from the Q DAC to the P DAC and the value after that loaded to the Q DAC. At the same time, the sawtooth waveforms are simultaneously reset, ready to draw the next vector. An alternative, slightly more economical, scheme uses two triangular waveforms displaced by half a cycle, at half the display dot sample rate. The successive sample values are loaded to the two DACs alternately, the result being the same as in the previous case.

The use of dot joining can avoid a hazard known as *perceptual aliasing*, illustrated in Figure 8.7. In (a), it appears that one cycle of a sinewave, in four different phases, is being displayed. With dot joining (b), it can be seen to be a much higher frequency waveform, the apparent variations in amplitude being due to there being only about four samples per cycle.

When deciding whether a digital storage oscilloscope is suitable for making a particular type of measurement, it is useful to have some quantitative guide with which to compare it with the

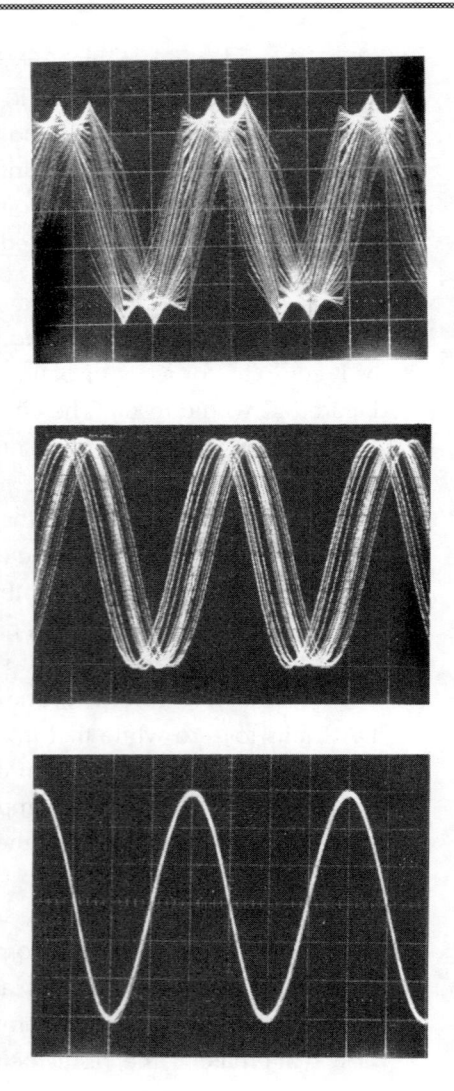

Digital storage oscilloscopes

Figure 8.5
The waveforms acquired by a digital scope are digitized under the control of a free running clock. When multiple acquisitions of a signal are stored, the timing relationship between the clock and the trigger can vary ±0.5 sample interval. Horizontal jitter is the result. Jitter is minimized by using large memories to store the waveform (which makes each horizontal element smaller), but the jitter will still limit horizontal magnification. Jitter correction is a feature of some digital storage scopes. Actually, any DSO that has random equivalent time capabilities has this feature built-in, as it is necessary to know where the sample points were sampled in relation to the trigger point.

Displays and display technology

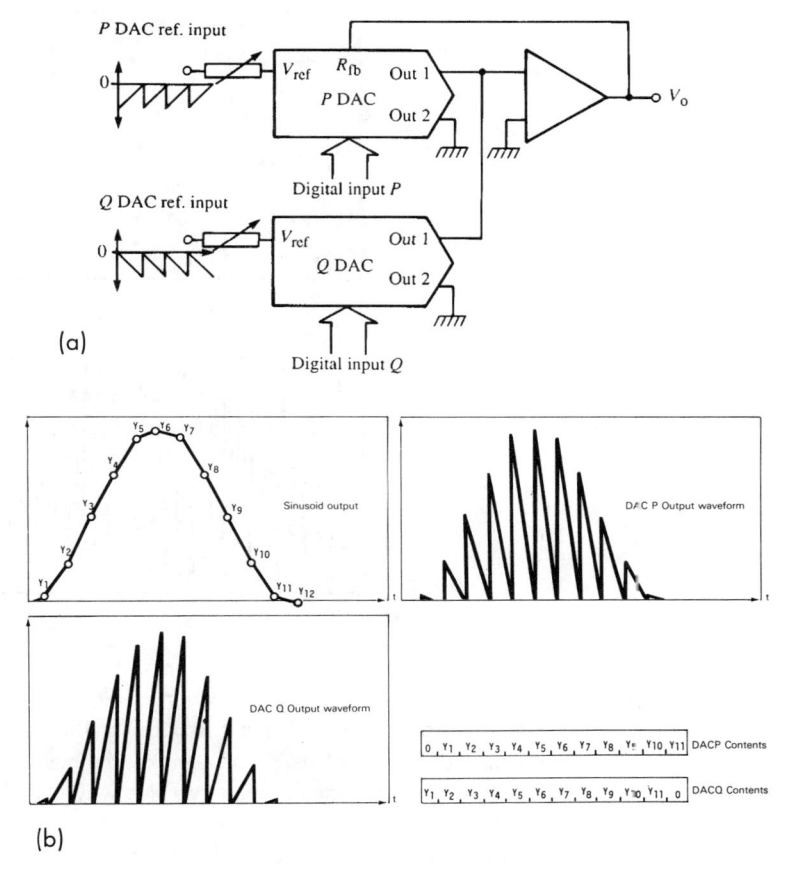

Figure 8.6
Interpolating DACs (a); waveform synthesis (b). (Both reproduced courtesy of *Electronic Engineering*.)

performance of a realtime analog oscilloscope. In the case of risetime measurements, the risetime of an analog scope, Tr in ns, is given in terms of its bandwidth BW in MHz by Tr = 0.35/BW, approximately, when an edge with a risetime very much shorter than this is applied. The measured risetime of an actual edge will be accurate to within 1% if the risetime of the edge is ten times longer than that of the oscilloscope. If comparable or shorter, the measured risetime will depend upon the risetimes of both the edge and the oscilloscope, so that:

$$Tr(measured) = \sqrt{\{(Tr_{scope})^2 + (Tr_{signal})^2\}}.$$

From this formula, if the risetime of the oscilloscope s known, the

77

Digital storage oscilloscopes

(a)

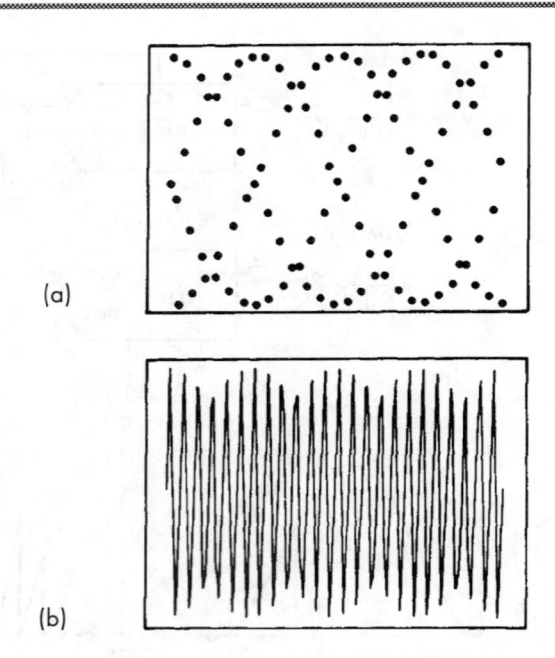

(b)

Figure 8.7
Perceptual aliasing errors are so named because sometimes the dot display can be interpreted as showing a signal of lower frequency than the input signal. But this is not true aliasing. The actual waveform is there; your eye – not the scope – makes the mistake. Note that in (a) what seems to be many untriggered sine waves is really one waveform. When vectors are drawn between the points in (b) note that vector displays can prevent perceptual distortion but can still show peak amplitude errors when data points do not fall on the signal peaks. (Reproduced by courtesy of Tektronix Inc.)

actual risetime of the signal can be back calculated from the measured risetime. For example, if the risetime of the oscilloscope equals that of the pulse being measured, the displayed risetime will be $\sqrt{2}$ times the true risetime.

The bandwidth is measured between the 10% and 90% points on the rising or falling edge, so that in the case of a digital storage oscilloscope using a pulse interpolator, when no samples fall on the edge the measured risetime will just be 80% of the sample interval, see Figure 8.11. Similarly, where just one sample falls within the range 10% to 80%, the measured risetime will be 1.6 times the sample period. The risetime measurement error may be defined as:

$$\text{error }(Tr) = (Td - Ti)/Td$$

where Td is the displayed risetime and Ti is the input signal risetime. If the latter is zero, the displayed risetime is due entirely to the oscilloscope, and the *displayed risetime error* is 100% of the sample interval. Thus the worst case risetime measurement for pulse interpolation reaches 100% when the actual risetime is as short as 1.6 sample intervals, though the actual error could be less than this. If one imagines an analog oscilloscope with a risetime of 1.6 times the sample period of a digital storage oscilloscope, it is possible to compare their performance. This shown in Figure 8.12, indicating that the worst case risetime measurement error for a digital storage oscilloscope is similar to that of comparable analog oscilloscope. But there is an important difference. In the case of the analog oscilloscope, knowing its risetime one can calculate the actual risetime of the signal from the measured risetime indicated by the oscilloscope. In the case of a digital storage oscilloscope, on the other hand, one knows only the worst case risetime, not the actual risetime — so it is not possible to deduce the signal's true risetime. This is just one of the many subtle differences between analog and digital storage oscilloscopes, which are commonly overlooked.

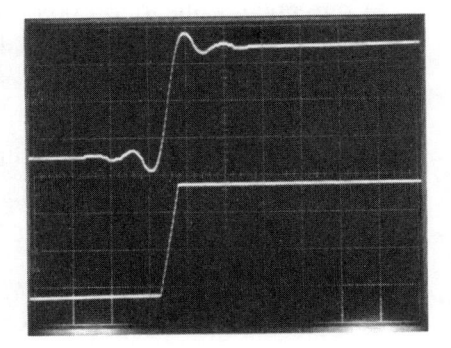

Figure 8.8
Displays constructed with sine interpolation avoid perceptual aliasing and envelope errors when used to display sine waves. But an interpolator designed for good sine wave response can add what appears as pre- and over-shooting to the display of a step function when there are less than three samples taken on the step. The error is minimized if more than three samples are taken and with narrow spectrum waveforms such as sine waves. The photo above is a double exposure of a signal with no samples on the step; the first trace is drawn with a sine interpolator and the second with a pulse interpolator. (Reproduced by courtesy of Tektronix Inc.)

A further improvement upon the dot-joining (pulse interpolator) display is provided by sine interpolation. This can provide a realistic representation of a sinewave with as few as 2.5 samples per cycle, see Figure 8.3, bottom row. The implication is that for the particular sine reconstruction algorithm used in this case, the useful bandwidth:

$$\text{USB(MHz)} = (\text{Maximum digitising rate in Ms/s})/2.5$$

This provides a useful yardstick for comparing the bandwidth of a digital storage oscilloscope with that of a realtime analog oscilloscope. Here again however, bear in mind that the illustration is of a single shot, or is as would appear if the 25MHz sample rate were exactly 2.5 times the frequency of the displayed 10MHz sinewave. If not, the display — although looking sinusoidal — will be subject to horizontal jitter as evident in Figure 8.5, centre photograph. (Note that all the displays in Figure 8.5 are greatly expanded in the horizontal or X direction.) The jitter is due to the fact that generally the frequency of the signal being acquired will bear no exact relation to the frequency of the sampling. Consequently, the trigger point which initiates the acquisition may occur immediately before the first sample, or up to one sample period before it, or any value in between. Some digital storage oscilloscopes are designed to measure this trigger-to-sample lead and adjust the horizontal position of the corresponding display appropriately, resulting in no apparent jitter even when displaying the waveform with X expansion, Figure 8.5, lower photograph.

While sine interpolation can work wonders if the waveform you are observing is sinusoidal, or at least smoothly varying, it is inappropriate for fast pulse waveforms, that is to say steps or pulses where, at the instrument's current timebase speed setting, there are two or less samples falling on the edge itself. With no samples falling on the step, a pulse interpolator will show a straight line representation of the rising (or falling) edge, joining one sample to the next (Figure 8.8, lower trace). However, the software reconstruction algorithm used by a sine interpolator operates upon a group of points, steadily advancing along the stored data. Consequently, when the group encounters a discontinuity (implying the presence of energy with frequency components above the Nyquist rate) it causes pre- and post-

Displays and display technology

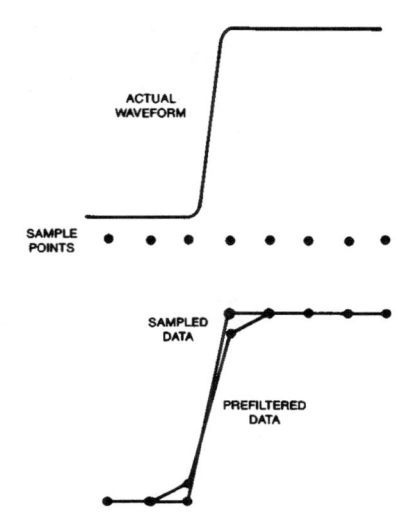

Figure 8.9
The pre-filter looks at the slope of three samples, then the next three, then checks for a discontinuity between the slopes. If there is a discontinuity of more than one division between the compared slopes, the closest points to the discontinuity are adjusted by about 10% of the amplitude, as shown. This waveform is then processed by the sine interpolator and, as the risetime has been effectively 'rolled off', the pulse is recontructed without the normal 'ringing' associated with sine interpolation. (Reproduced by courtesy of Tektronix Inc.)

ringing to be displayed (Figure 8.8, upper trace).

In addition to the obvious waveform distortion, a sine interpolator misrepresents a fast rising edge in another way. With only a few samples falling on a step, sine interpolators tend to show a risetime faster than it really is, due to the ringing introduced on the edge. Some oscilloscopes have a software digital data prefiltering algorithm which can be used in conjunction with sine interpolation when the waveform being investigated is not necessarily sinusoidal. When this algorithm detects a gross value change from one sample to the next, the values are modified before being sent to the sine interpolator algorithm (Figure 8.9).

The result is reduced ringing introduced when displaying fast edges, as shown in Figure 8.10.

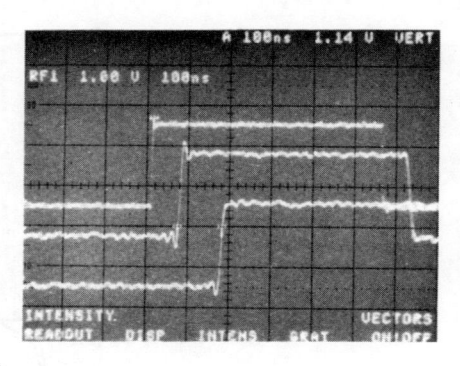

Digital storage oscilloscopes

Figure 8.10
The top pulse in the photo is a pulse acquired in equivalent time mode which represents the actual waveform. The middle pulse is the same pulse acquired in the sine interpolate mode with the prefilter off; the same pulse is shown at the bottom with the prefilter on. The filter helps to make the display more representative of the waveform. (Reproduced by courtesy of Tektronix Inc.)

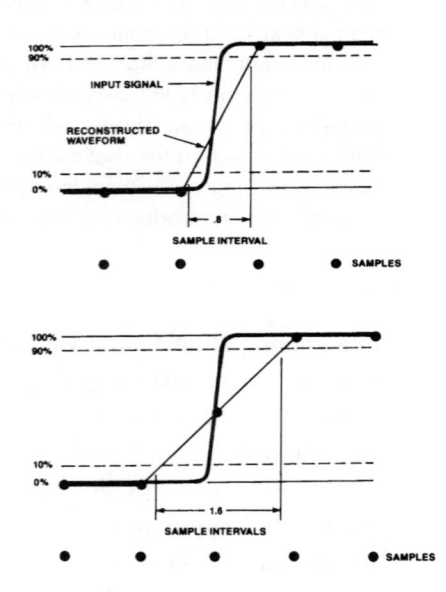

Figure 8.11
To demonstrate how the errors in a rise time measurement made by a digital system can change with the sample placements, the same input step is shown in both drawings. In the first, the step occurs exactly halfway between samples. The rise time of the resulting vector display is 0.8 sample intervals in this situation. On a different acquisition of the same signal, however, the samples may fall as shown in the second drawing. In this worst case, the rise time indicated by the display is 1.6 sample intervals – the maximum possible. (Reproduced by courtesy of Tektronix Inc.)

Displays and display technology

Figure 8.12
A computer model of a digital storage scope display was used to generate the rise time error ranges shown here. To make the results independent of any particular digitizing rate, the horizontal axis is the number of sample points per displayed rise time on the input step function, while the errors plotted vertically are shown in percentages of displayed rise time. The input step was a worst case — an exponential step. For comparison, the error curve of an analog system with an equivalent rise time is also plotted. (Reproduced by courtesy of Tektronix Inc.)

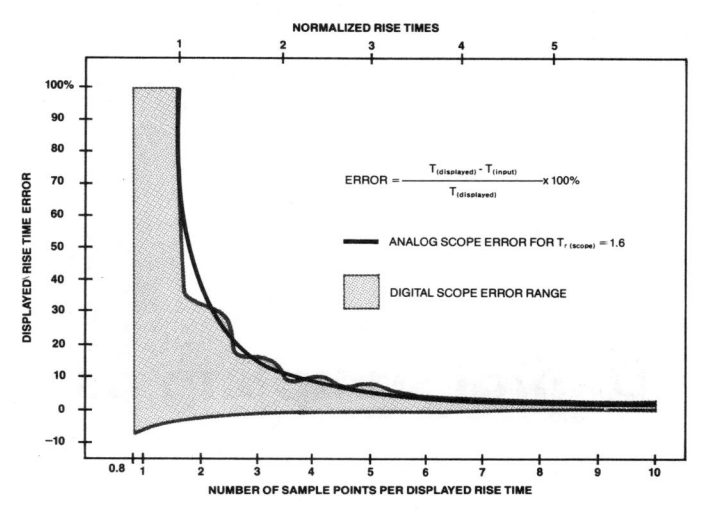

Chapter 9

Digital sampling oscilloscopes

The maximum sampling rate of digital storage oscilloscopes has increased enormously over the years, from a few hundred kilosamples per second to 5Gs/s, which at the time of writing currently represents the state of the art. This performance enables a user to capture a waveform with significant energy at frequency components up to 1GHz, on a single shot basis, or a sinewave at the full 1GHz bandwidth of the input circuitry on an oscilloscope, without using equivalent time sampling and therefore with no danger of a misleading display due to aliasing. A good example of this advanced type of digital storage oscilloscope is shown in Figure 9.1. Thus digital storage oscilloscopes now rival the fastest analog oscilloscopes, such as the legendary Tektronix model 7104, which was the fastest oscilloscope in the world in its day.

But even that oscilloscope's bandwidth was surpassed by a factor of two by an oscilloscope which appeared as early as the late 1950s. A 2GHz bandwidth at that time, when other oscilloscopes struggled to make a bandwidth of 85MHz, was almost unbelievable. But the Hewlett-Packard model 280 (I think, from memory, that was the model number) was a sampling oscilloscope. Subsequent developments extended the bandwidth of sampling oscilloscopes to 14GHz by the early 1970s, while today's digital sampling oscilloscopes, such as the 11801B from

Digital sampling oscilloscopes

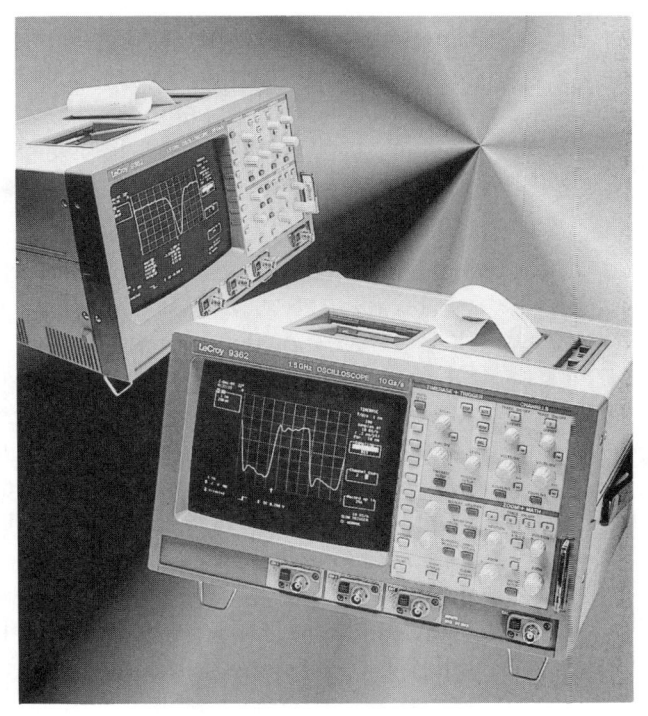

Figure 9.1
The LeCroy model 9362 features a 5Gs/s sampling rate simultaneously on its two input channels, providing a 750MHz single shot bandwidth and a 1.5GHz equivalent time bandwidth for repetitive signals. In single channel operation, the instrument can sample at 10Gs/s. Vertical resolution is 8 bits, or up to 11 bits with optional digital low-pass filtering. (Reproduced by courtesy of LeCroy Corporation.)

Tektronix, offer a bandwidth of no less than 50GHz.

So what is a sampling oscilloscope, and how does it work? And how does a digital sampling oscilloscope differ from it? Answering the second question first, there is very little difference: in a digital sampling oscilloscope, the samples taken are digitised, which enables them to be stored in semiconductor memory. In contrast, in the original type of sampling oscilloscope, now called an analog sampling oscilloscope to distinguish it from the digital variety, the samples were simply displayed on the screen of a cathode ray tube as they were taken. The only 'memory' was provided by the persistence of vision, the sampling and display rate being adequate to provide a steady, flicker free picture of the waveform under investigation. The similarity between the earlier sampling oscilloscope and the modern digital sampling oscilloscope is, indeed, so great, that the greater part of this chapter is taken up with describing how the basic original sampling oscilloscope works. The digital sampling oscilloscope is virtually the same animal, but with provision for digitising the samples.

Sampling oscilloscopes, both digital and analogue, achieve their great bandwidth by bypassing all the bandwidth limiting elements of a conventional realtime analog oscilloscope; the input attenuator, Y amplifier, Y deflection stage and of course the cathode ray tube itself. The sampling oscilloscope avoids all these limitations at one fell swoop, by not dealing with the whole signal in real time. Instead, it takes samples of the instantaneous voltage of the input signal on successive cycles and assembles these sample to form a picture of the complete waveform. At first sight, this sounds like what a digital storage oscilloscope does in equivalent time mode: the crucial difference is in how the samples are taken. As with the equivalent time mode of a digital storage oscilloscope, a sampling oscilloscope can only operate in this way if the signal goes on repeating from cycle to cycle for a long as it takes to build up the display. Hence the sampling oscilloscope is limited to displaying repetitive waveforms. This is one limitation. Another results from the omission of input attenuator and input amplifier. The size of the largest signal a sampling oscilloscope can handle is quite restricted, only a volt or two peak to peak — including any d.c. component. Often, when using a sampling oscilloscope, the user is interested only in the a.c. behaviour of the circuit under investigation. So a.c. coupling can be used to prevent any d.c. level present eating away at the usable peak to peak input voltage range, while for handling larger signals, a X10 or X100 probe tip attenuator can be used. Likewise, the omission of an input amplifier limits the usable range in the other direction, the smallest signal swing viewable being limited by sampling noise — the inevitable small sample-to-sample voltage variations which occur even when the input voltage itself is not varying.

Thus the main requirement for a sampling oscilloscope is a circuit capable of accurately sampling the input waveform at regular intervals. In a nutshell, this is the stroboscopic technique used to

Digital sampling oscilloscopes

slow down the motion (or frequency) of events which are too fast to observe by conventional means. If one wants to study some mechanical event like the turning of gears which rotate too fast for the eye to see, they can be illuminated by a stroboscope. If they are repeatedly briefly lit once per revolution, or once after several complete revolutions, they will present a stationary image. But if after each revolution (or group of revolutions) they are lit up a small amount of time later (say Δt later), then the eye sees samples of successively later positions, and if this happens continuously, the eye is deceived into seeing continuous (albeit much slowed down) motion — the same effect which in a movie makes the spokes of a wheel seem to be turning slowly or even turning backwards when the vehicle is in fact travelling rapidly forward. This is a direct analogy of *sequential sampling*, the normal operating mode in sampling oscilloscopes.

Figure 9 2. illustrates the basic process. A signal is applied to the

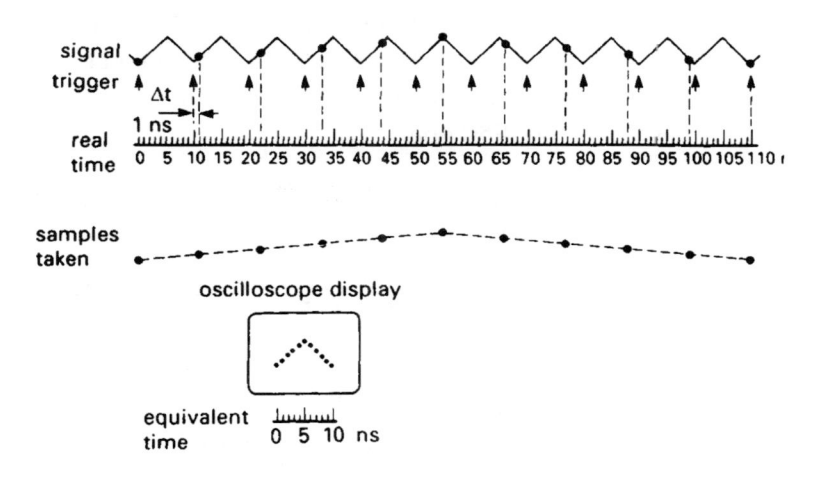

Figure 9.2
A sampling oscilloscope uses sequential sampling, illustrated here.

vertical input of the sampling oscilloscope, and also (internally or externally) to its trigger circuitry. Assume that the negative tip of the waveform just causes triggering, and that the first sample of the signal voltage is taken at that instant.

On the oscilloscope screen, a dot appears at the correct vertical

Digital storage oscilloscopes

position. On the next signal cycle, a trigger pulse again occurs at the same point on the waveform, but this time, circuitry in the oscilloscope delays the taking of the sample by the time increment Δt. This second dot will appear at an appropriately higher level on the screen of the cathode ray tube, and it will also be displayed displaced to the right by a distance representing the time delay Δt. The timescale on the oscilloscope screen does not represent, as in ordinary real time oscilloscopes, the actual or real time at which the sample was taken (11ns after the first sample), but represents instead the time equivalent to the distance between the two samples had they occurred on one and the same signal cycle (1ns). The user of a sampling oscilloscope will not usually be aware of, nor want to know, how much real time elapses between samples; he or she is only concerned with the timescale of the reconstructed image. In practice several or many complete cycles may elapse between the taking of one sample and the next. In Figure 9.3. the number of cycles *n* elapsing between samples is shown as five, but the number *n* could be very much greater, it is of no concern to the user of a sampling oscilloscope.

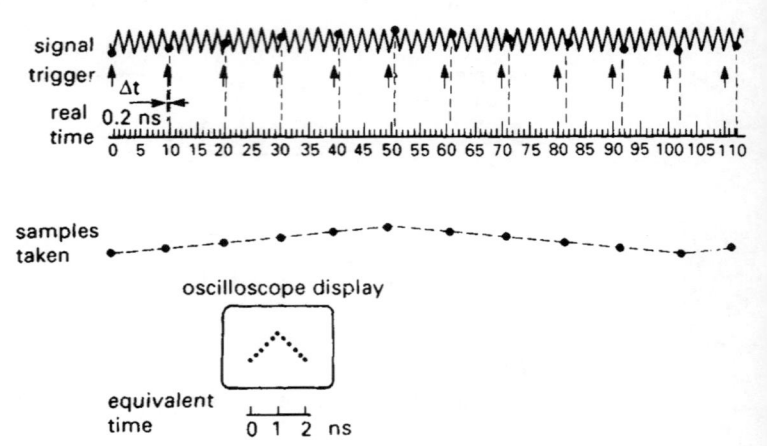

Figure 9.3
A sampling oscilloscope may take one sample per cycle of the input, as in Figure 9b, or only one sample per *n* cycles of the input, as here. In this example *n* = 5.

Digital sampling oscilloscopes

To briefly recapitulate. The screen display is built up of discrete dots whose vertical positions correspond to the signal voltage at the time of sampling and whose horizontal positions correspond to the total time delay between the beginning of the waveshape and the moment when the sample was taken. To recognize the beginning of the waveshape, the instrument uses trigger circuitry much like that of a conventional real time oscilloscope. For triggering from very high frequency signals, e.g. from several GHz upwards, the trigger circuitry may be aided by a prececing divider stage. This may employ a tunnel diode or other special circuitry, providing *trigger countdown*.

The ability of the instrument to recognise a certain point on the waveform and trigger on it and then delay the taking of successive samples by increasing amounts of time means that the sequential sampling oscilloscope can do something which no stroboscope can do: it can successfully sample a repetitive but irregularly occurring waveform. This is illustrated in Figure 9.4. Note, however, that this capability is limited to sequential sampling mode. The random sampling mode will not work with irregularly spaced samples, for reasons which will become apparent later. Therefore, if it is desired to view the leading edge of an irregularly occurring fast pulse, a delay line must be used between the trigger pick-off point and the sampling gate as indicated in Figure 9.5.

The sequential sampling oscilloscope

The block and timing diagrams of a typical analog sampling oscilloscope, operating in sequential mode, are shown in Figure 9.5. The signal is routed from the input socket to the trigger take-off circuit, where a few percent of the signal energy is extracted for use in the internal trigger mode (if provided). Alternatively, an external trigger signal can be used, if available. In this case, and if a delay line is not used, the sampling gate may be mounted in the business end of a probe not unlike, in size and shape, the familiar 10:1 passive divider probe used with ordinary oscilloscopes. Probe type sampling heads offer bandwidths up to a few GHz combined with a high input impedance — typically 100KΩ in parallel with 0.5pF. High impedance sampling heads become impractical at frequencies much higher than this, so really high frequency sampling heads, with bandwidths up to

Digital storage oscilloscopes

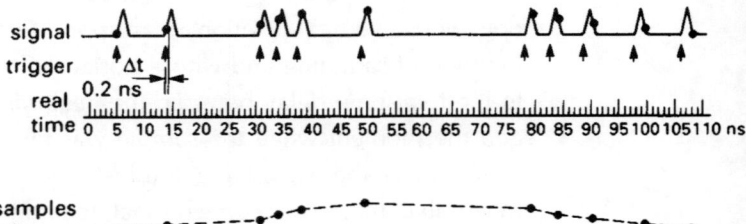

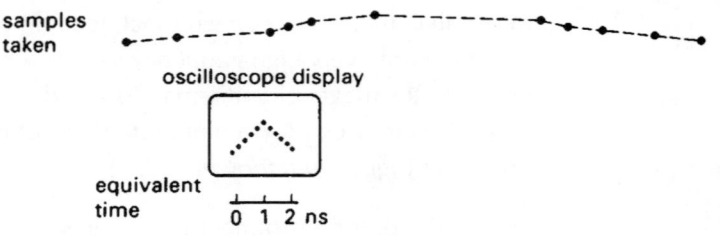

Figure 9.4
A sampling oscilloscope can, under favourable circumstances, capture a repetitive waveform, even though the prf (pulse repetition frequency) is not constant.

50GHz, have an input impedance of 50Ω. In this case, the sampling head is usually accommodated within the oscilloscope, often within a plug-in unit, the signal of interest being connected to the oscilloscope via a length of high quality coaxial cable.

Trigger and hold-off circuitry is comparable to that in any high frequency real time oscilloscope (preceded by trigger count-down for operation at the highest frequencies), and if the trigger level and slope are correctly set, the trigger pulse, waveform 1 in Figure 9.5, can be made to occur very near the beginning of the waveshape of interest.

The two blocks following the trigger circuit provide the variable delay. Each trigger starts a fast ramp, waveform 2. This ramp and waveform 3 are fed to the strobe comparator, and when the ramp has run down to the initial level of waveform 3 the comparator puts out a sampling pulse. This pulse operates the sampling gate which consists of a set of balanced diodes (the exact configuration of which varies — depending upon the particular make and model) which are made to conduct, and so permit the signal voltage present on the input side to appear at the output of the gate. No matter how low the source impedance of the signal is,

Digital sampling oscilloscopes

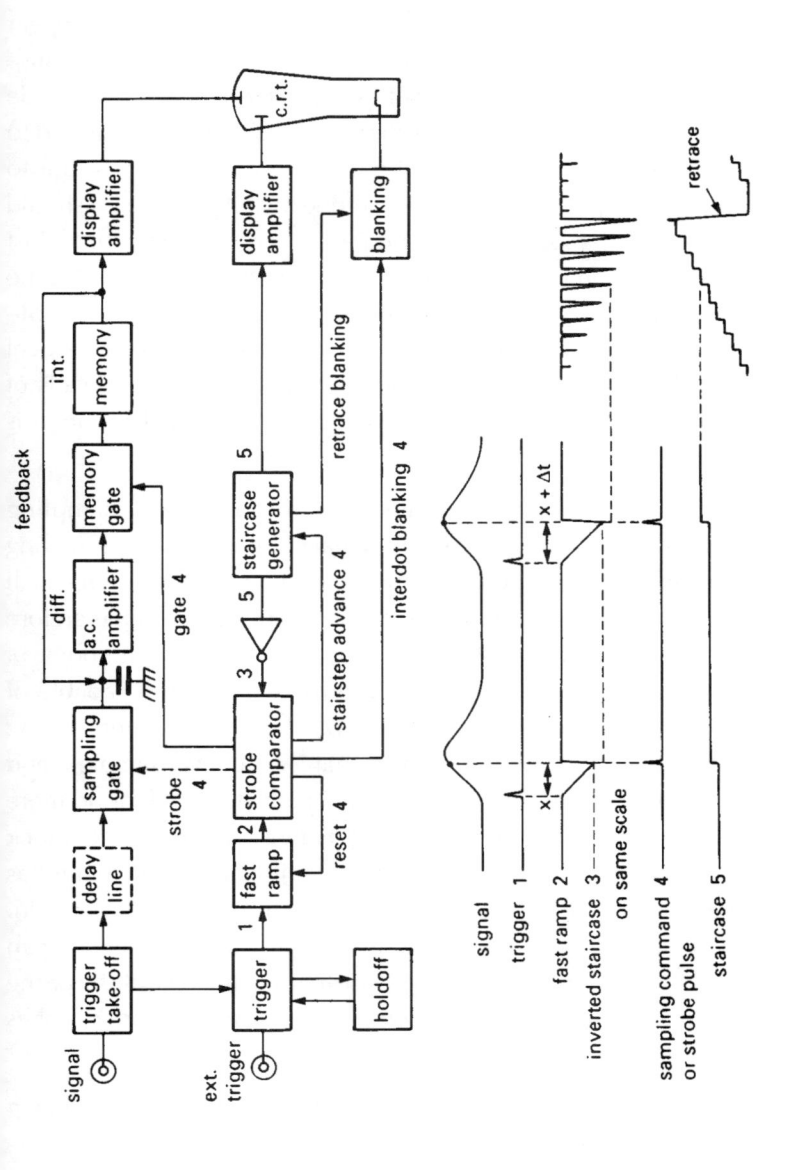

Figure 9.5
Block diagram of an analog sampling oscilloscope, showing the waveforms observed at the points indicated.

Digital storage oscilloscopes

and how high the input impedance of the sampling gate, some finite amount of energy will have to be drawn from the signal circuit during the gate conduction period to charge the capacitor at its output. To this extent, the gate is interfering with the signal, and at frequencies in the GHz region this interference will appear as a waveform distortion known as *kickout*. In order to minimise it, the capacitor at the output of the sampling gate is made extremely small, which means that the voltage level transferred to it will disappear in a matter of nano- if not picoseconds. But to produce a bright display on the cathode ray tube the beam should be held at the point corresponding to each sample until just before the circuit is ready to take the next sample. The level at the output of the sampling gate is, therefore, after suitable amplification, gated into a memory which then drives the vertical deflection plates via a conventional output amplifier. (Details of the feedback loop shown in Figure 9.5. follow later in this section.)

As the fastest sampling rate used in a typical sampling oscilloscope is, say, 200kHz the signal coming from the sampling gate cannot possibly change any faster than once every 5μs. All the oscilloscope circuits following the sampling gate can therefore be designed in the most modest way. This is the whole point of sampling. One cannot design real-time oscilloscopes capable of looking at signals of many GHz, or risetimes of a few picoseconds. But one can, if the signal is repetitive, sample it and handle the sample gate output with circuits designed for no more than, say, 1MHz. (By comparing Figure 9.5. with the block diagram of a conventional oscilloscope, it can be seen that it has disposed of all the bandwidth limiting items in the Y signal chain. The limiting factor now is how short a sample of the signal it can take — if a wheel is turning so fast that each single flash of the stroboscope illuminates one complete revolution, shorter flashes are needed.)

Waveforms 2 and 3 in Figure 9.5. have resulted in a sample being taken as shown, at a particular point on the signal waveform. The circuit must now readjust itself so that after the next trigger the sample is taken later (relative to the signal cycle) by an amount Δt. To do this, the level of waveform 3 is adjusted as shown, and this can in fact be initiated by the strobe pulse. The succession of

Digital sampling oscilloscopes

small d.c. levels builds up to a waveform known as the staircase, and since each step of this staircase corresponds to a time increment Δt, and we wish to move the cathode ray tube beam horizontally by an amount corresponding to these same time increments, the staircase waveform can also be used to drive the horizontal deflection circuits. The fact that in Figure 9.5. the staircase 5 is a positive-going waveform and the strobe comparator requires a negative-going waveform is of no deep significance. The designer could have chosen a positive-going fast ramp for 2, and then 3 could be replaced by 5, saving the inverter. It has been shown as in Figure 9.5 simply because that is the way the majority of actual analog sampling oscilloscopes worked. Note that the staircase waveform could be replaced by a ramp generator. However, in this case, we would be limited to sampling regularly recurring, jitter-free wave forms as in Figure 9.3. By using a staircase, where the next step is initiated by the sampling command, it is possible to sample an irregularly occurring pulse, as in Figure 9.4, since the 'treads' of the steps need not all be the same width. However, this clearly only works if the shape of the signal pulse is constant.

To the right of the detailed waveforms, 2 and 5 have been redrawn to a compressed timescale to show the complete sequence. With a sufficient number of samples across the screen, the staircase will be made up of so many small steps that to the naked eye it appears exactly like a conventional sweep sawtooth, in the same way that in the sampling display itself the dots merge to give the appearance of a continuous trace. Blanking is used not only to prevent the appearance of the retrace or flyback, but also to cut off the cathode ray tube while the beam moves from one discrete sample position to the next; this is known as *interdot blanking*.

Now to return to the feedback loop in the vertical circuit, and the reason for including it. It was stated earlier that energy is drawn from the signal circuit to charge a small capacitor (typically just the wiring and stray capacitance) at the output of the sampling gate. When the gate stops conducting, the voltage to which the capacitor was charged will quickly leak away: but before this can happen it is amplified and gated into a memory circuit which will hold this level. Without any further circuit complications, the memory could be reset shortly before the next sample is due to be taken and the process started all over again. Such open loop memories were used in some cheaper types of early analog sampling oscilloscopes.

The advantage of introducing a feedback loop is two fold. First, the feedback can be used to hold the voltage of the capacitor at the output of the sampling gate at the level of the sample just taken, and then if the signal, when the next sample is taken, happens to sit at the same voltage level, no energy need be drawn from it: the gate output circuitry is already at that level. This minimises kickout. It might seem a surprising assumption that the signal level might be the same on successive samples, but if a sufficient number of samples are taken to create a reconstituted display where the individual dots merge into a continuous trace, this does in fact mean that the signal level voltage-changes from sample to sample are very small. The second advantage of the feedback loop is that it is self-correcting, making the circuit performance nearly independent of amplifier gain variations.

Figure 9.6. illustrates how the loop works. The diagram shows the situation where the signal, when it is first sampled, is two units high. But the sampling pulse is extremely narrow (as short as 30ps or less) for the reason indicated earlier, and does not give the capacitor time to charge to the full two units. In Figure 9.6 it is shown as charging to only half a unit. This represents a *sampling efficiency* of 25% — actually an optimistic assumption. (Typical values might even be as low as two percent.) The a.c. amplifier is a slowly responding one with the aim of getting from it an amplified and 'time stretched' version of the input. The memory gate pulses, although initiated from the strobe comparator, are also made comparatively long (typically 300ns). The memory is acting as an integrator, and its output is the cumulative result of successive inputs. This output drives the cathode ray tube display amplifier. It is also fed back to the input via a very slow time constant network where it will take nearly 5µs to raise the input capacitor of the ac amplifier to the level that the signal had when the sample was taken. The slow time constant explains why the sampling system cannot take samples faster than at a 200kHz rate.

Why use such a slow time constant? As can be seen, the feedback

Digital sampling oscilloscopes

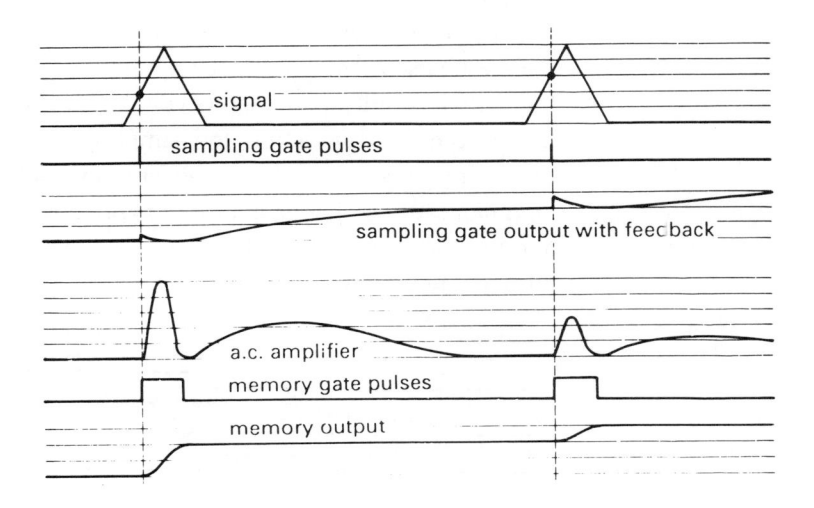

Figure 9.6
The small output from the sample gate is amplified by the a.c. amplifier up to the level it would be if the sample gate efficiency were 100%, and gated into the memory. This level is then fed back, jacking up the dc reference level of the a.c. amplifier input to the 100% level.

used is in fact positive feedback (in the same direction as the original signal) and if it arrived while the memory gate (not the sampling gate) was still conducting, it would lead to oscillation.

The gradual raising of the voltage on the input capacitor to the correct level of two vertical units is of course also amplified by the a.c. amplifier, which explains the second, longer, lower bulge in its output waveform. But as the memory gate is not conducting during this period, it is of no significance. It is worth noting that the combination of a.c. amplifier (acting as a differentiator) and memory (acting as an integrator) ensures that the d.c. component of the signal will in fact be passed by the circuit.

Figure 9.6 shows a second sample then being taken, and because at this time the signal is at three vertical units and the sampling gate output already sits at two vertical units, the circuit sees a potential difference across the gate of only one unit. With a sampling efficiency of 25%, the output moves only a quarter of a unit before the sampling pulse ends, but with the same circuit gains as before this results in just the right amount of change to bring the memory output to the correct level.

Digital storage oscilloscopes

Looking now at the solid-line drawing of Figure 9.7, the more common case is shown where, at the time of the second sample, the signal is still at the same voltage as on the first. There is therefore no voltage across the sampling gate when it conducts, no energy need be transferred, no kickout occurs, the a.c. amplifier sees no change at its input and thus produces no output, and the memory remains the same level. All is well in the best of all possible worlds.

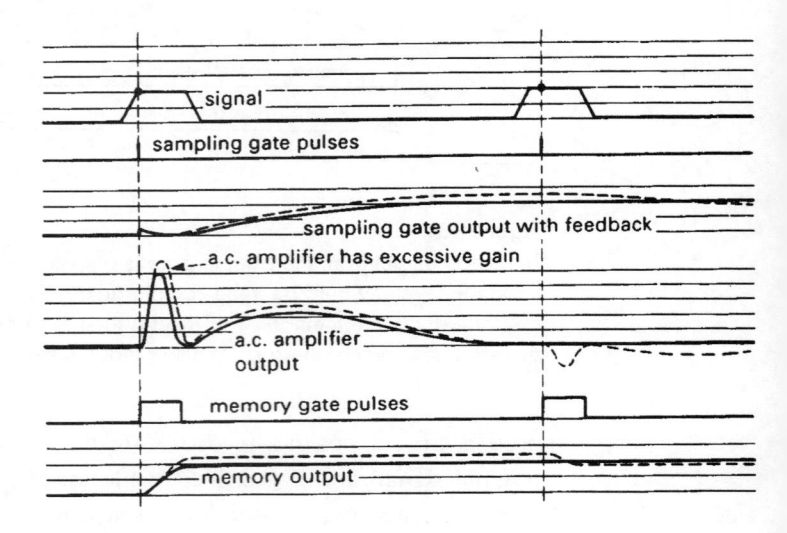

Figure 9.7
Although the memory feedback is positive, overall the loop shows the self-correcting action of a negative feedback loop.

But Figure 9.7 also illustrates with dashed lines how the feedback loop takes care of departures from this ideal. As an example, it has been assumed that the a.c. amplifier gain is excessive. This means that the memory output will be too high, and the dot will appear too high on the cathode ray tube. Because, in Figure 9.7, the signal level for the second sample is unchanged, the action of the feedback loop can be seen very readily. When this second sample is taken, the voltage at the gate output is in fact (erroneously) too high, so energy will be transferred in the opposite direction and the gate output voltage will drop down (by the usual 25% of the difference). This negative change is seen and amplified and added to the memory, but as the a.c. amplifier gain is excessive, it will again result in too much movement. The original overshoot is over-corrected, giving an undershoot of small amplitude. On the third sample the overshoot is reduced still further and on successive samples the circuit rapidly settles to the correct level. If the samples were widely spaced and individually discernible the appearance would be like that of a damped oscillation (see Figure 9.8 (a)). Thus while in the short term, i.e. on a sample-by-sample basis, the feedback loop provides positive feedback, in the long run it demonstrates the self-correcting, distortion-reducing effects of negative feedback.

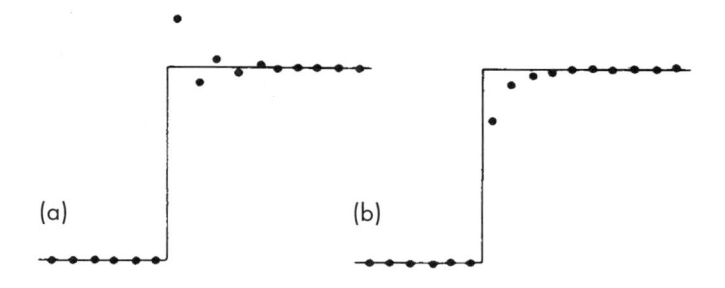

Figure 9.8
Illustrating the self-correcting action of the sample loop for conditions of high (a) and low (b) loop gain. Many digital sampling oscilloscopes include self-calibration routines to automatically set up the correct loop gain.

Exactly the same effect would occur if, instead of excessive a.c. amplifier gain, the memory circuit had too much gain, the feedback path had less attenuation, or the sampling efficiency increased. All these conditions are covered by expressions like 'the sampling system has too much loop gain'.

If, conversely, the a.c. amplifier gain had been too low, the first sample would not have reached the correct level, and the difference between the ideal and the actual level would have again be seen by the circuit when the subsequent samples were taken. In this case, the result is a gradual approximation to the correct level, giving the appearance of simple undershoot (Figure 9.8 (b)), a condition known as *low loop gain*.

This section has described how, in a traditional analog sampling oscilloscope, positive feedback is used to boost the sampling gate output from just a few percent to effectively 100%, enabling the true signal amplitude to be measured at each sample. A similar scheme is used in digital sampling oscilloscopes, but in these, the voltage across the short term integrator 'memory' (in practice, a capacitor) following the memory gate in Figure 9.5 is digitised and stored. This represents one type of sample feedback loop (SFBL), the analog variety. Alternatively, the signal sample can be digitised in an ADC directly and stored in memory, the SFBL being closed by means of a voltage derived from a DAC fed with the sample value just taken, or in some instruments, with the corresponding sample value stored on the previous acquisition in repetitive mode. Both analog and digital SFBLs are used by different manufacturers of digital sampling oscilloscopes, one major manufacturer using both types, in different models. A detailed discussion of the relative merits of these two schemes is beyond the scope (no pun intended) of this book. However, it is worth noting that with an ADC directly at the memory gate output (see Figure 9.5), there is no possibility of droop in the feedback voltage with very slow sampling rate. But a very high resolution ADC (and associated DAC) is needed to cope with small signals. Whereas with an analog SFBL, the different Y sensitivity ranges can be provided by a switched gain amplifier between the integrating memory output and the ADC.

Another feature of digital sampling oscilloscopes which one should be aware of, is *blow-by*. Kick-out — the appearance in the circuit being monitored of disturbances caused by the action of the sampling head — has already been mentioned. Blow-by is the other side of the coin, being the result of leakage of the signal voltage through the sampling gate (see Figure 9.5) when it is not conducting. The isolation of the gate when in the OFF condition (i.e. most of the time) cannot be made totally complete. So the voltage at the input of the a.c. amplifier when the next sample is about to be taken will be (incorrectly) influenced marginally by what the signal voltage did between the last sample and now. (Note that the sample feedback voltage does not in fact go to the small temporary storage capacitor at the input of the a.c. amplifier directly, but via a high resistance, so that when the sampling gate

Digital sampling oscilloscopes

is ON, it does not 'see' the output impedance of the integrating memory.)

It is impossible to design a sampling gate with infinite attenuation in the OFF state, and different ways have been developed of dealing with this problem. In a simple two diode sampling gate, such as that shown in Figure 9.9, the blow-by is too large to ignore. In one model of digital storage oscilloscope, the blow-by is measured — by acquiring a full scan's worth of data with the sampling pulses absent and the sample gate therefore permanently OFF — and the signal then acquired on a second scan. On this second scan, each measured data point is corrected for blow-by by subtracting the value stored in this location on the previous scan. This corrected value is then stored in acquisition memory in place of the previously held blow-by value. This scheme is elegant, cancelling the effect of blow-by completely, at least in principle. The practical disadvantage is that two complete acquisitions are needed to obtain one valid picture of the waveform being monitored, making operation comparatively slow. In other models of digital sampling oscilloscope, a more complicated sampling gate is used. A six diode travelling-wave sampling gate is capable of providing adequate isolation when OFF to reduce blow-by to negligible proportions, obviating the need for two scans to obtain a single effective acquisition.

Sequential sampling oscilloscope behaviour

It was mentioned in the last section that the results of incorrect loop gain, and the action of the feedback loop in such cases was particularly well illustrated by Figures 9.7 and 9.8 because the signal level on subsequent samples was unchanged. Now on some analog sampling oscilloscopes a front panel control (usually labelled *dot response*) allowed the precise adjustment of the loop gain, and obviously the best kind of waveform to use during the adjustment is one resembling Figure 9.8 such as a squarewave. This facility was useful, as it also permitted the implementation of a technique called *smoothing*, which was available on most analog sampling oscilloscopes. This deliberately reduced the loop gain to a low figure, say one-third of normal. The result is that the first sample rises to only one-third of the final signal amplitude, and if the signal level remains unchanged, subsequent samples

Digital storage oscilloscopes

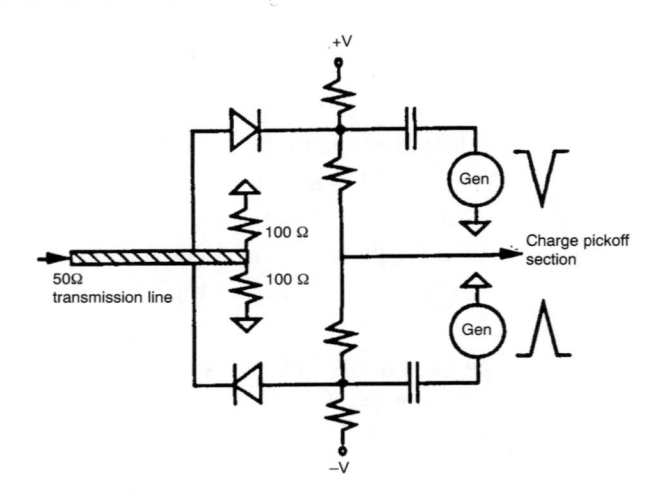

Figure 9.9
Simplified circuit diagram of a two-diode high speed sampling gate. The gate is opened for a few picoseconds only, in order to provide the necessary time resolution to sample signals of tens of GHz.
Consequently, its output (the input to the a.c. amplifier), reaches only a few percent of the full value. The error sample feedback loop makes up the deficit, in time for the next sample taken.

will each rise by one-third of the remaining difference, giving the usual exponential approach to the correct level. This is shown in Figure 9.10 (a). It can be seen that with a loop gain of one-third it takes twelve samples for the display to reach a value within 1 per cent of the final value. The smoothing technique reduced noise to a value corresponding to the reduction in loop gain: in the above example with a loop gain of one third, the noise is reduced by 9.5 dB. The important point is that if the dot density is sufficient, this noise reduction can be achieved without affecting the shape of the signal — in other words, without reducing the bandwidth of the system.

A price had to be paid, of course, and in this case the noise reduction was bought at the expense of time. With the great dot density needed in this mode, a flickering display or even a slow-moving spot resulted. Nevertheless, it shows how a technique which is easy in the digital world was in the past achieved in the purely analogue sampling scope. In a modern digital sampling scope, the same effect is achieved by acquiring the trace many

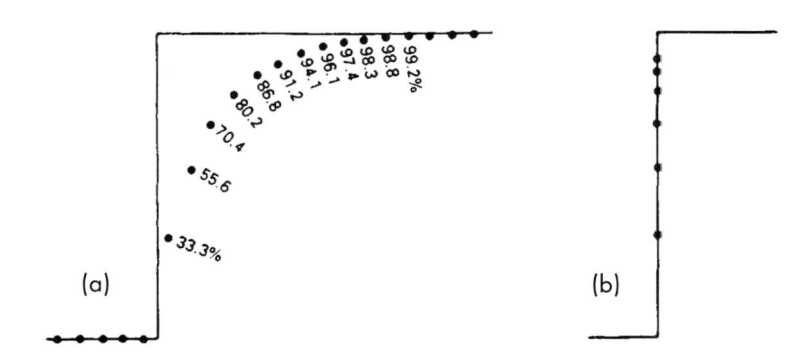

Digital sampling oscilloscopes

Figure 9.10
By deliberately reducing the sampling efficiency to, say, one third, the effect of sample noise is also cut by a factor of three, but many samples are needed to completely record the risetime (a). By sufficiently increasing the dot density (reducing the Dt sample interval (b), the risetime is not impaired: a measure of noise reduction has been achieved without loss of bandwidth. Whilst this facility was common in analog sampling oscilloscopes, digital sampling oscilloscopes achieve the same end by averaging successive traces.

times over and storing the average of the samples for each given point as the result for that point. Here again, the noise reduction is bought an the cost of increased time.

Noise reduction through averaging is only one of the many refinements included in modern digital sampling oscilloscopes. These instruments provide the highest bandwidth/shortest risetime of any type of oscilloscope, although as noted, their use is confined to recurrent waveforms. Lacking a single shot acquisition capability, it is clear that they cannot be usefully employed for troubleshooting the more elusive type of faults, such as when hunting an elusive glitch in a digital system.

A feature of some analog sampling oscilloscopes was a random sampling facility. This should not be confused with random repetitive (random interleaved) sampling in a digital storage oscilloscope. In fact, it was incorporated for a very good reason, namely the inability of a sampling oscilloscope, operating in the usual sequential mode, to see the leading edge of the trigger pulse. In sequential sampling mode, the leading edge could be seen, but only with the aid of a signal delay line. A sample of the

Digital storage oscilloscopes

signal, for triggering, was picked off ahead of the line, and the delayed signal at the line's output applied to the oscilloscope input. However, this arrangement had two drawbacks. Firstly, the limited bandwidth of the delay lines then available restricted the usable bandwidth of the sampling oscilloscope to well below its full potential, and secondly the user was faced with an input impedance of 50Ω, instead of the high impedance of a probe.

Random sampling enabled the leading edge of the signal, used to trigger the oscilloscope, to be seen, using a high impedance probe and no delay line. The way this worked is described in detail in *Oscilloscopes — How to use them — How they work* 4th Edition (ISBN 0 7506 2282 2), so only the briefest resumé is given here. In effect, the oscilloscope produces a repetitive internal trigger signal which is phase-locked to the repetition rate of the signal. Samples are then taken nominally centred on this trigger signal, giving a 50% pretrigger display. It is clear that the arrangement can only work with repetitive signals with low jitter, and not with irregularly occurring signals such as in Figure 9.4.

The scheme involves considerable circuit complexity and hence cost, and does not appear to be available on any current digital sampling oscilloscope. This is because the need for it has been largely overtaken by events, in the form of wider bandwidth delay lines, and wideband FET input active probes capable of driving a 50Ω load. For example, the Tektronix TDS820 Digitising Oscilloscope comes with built-in delay lines permitting viewing of the trigger event, and provides a 6GHz bandwidth. The instrument comes as standard with two P6207 FET probes providing 3.5GHz bandwidth at the probe tip, combined with an input impedance of 100K in parallel with less than 0.4pF. As it does not rely on random sampling, it can view irregular repetitive signals, such as radar responses.

Chapter 10

Accessories for use with digital storage oscilloscopes

There are many accessories available for use with oscilloscopes. Most of these were originally designed for use with analog oscilloscopes, but are equally of use with digital storage oscilloscopes. There is just one major exception, oscilloscope cameras. For while one of these can always be used to record the screen display, virtually all digital storage oscilloscopes (and even digital storage adapters such as that illustrated in Figure 7.2) have a built-in interface to allow the screen display to be dumped out on a printer (Epson quad density graphics being a common format for dot matrix printers), or downloaded to a PC over RS232, Centronics or HPIB interface. Nevertheless, some workers prefer a photograph of the screen display for use in reports, as somehow more convincing than a print-out, so oscilloscope cameras are occasionally used (quite unnecessarily, many would argue) with digital storage oscilloscopes.

Probes and probing

The first accessory to be considered is the 10:1 divider passive probe, which is used with any oscilloscope — analog or digital. Newcomers to the use of oscilloscopes do not always see the need for such a probe, especially as it reduces the sensitivity of the oscilloscope by a factor of ten — surely you just connect the circuit under test to the oscilloscope, and observe the waveform.

Digital storage oscilloscopes

Actually, it is not quite that simple. Although the Y input of an oscilloscope has a very high input impedance, in many cases its effect upon the circuit to which it is connected is not entirely negligible. The standard Y input resistance is 1MW and the input capacitance is usually in the range 15–30pF depending upon the particular make and model. With such a high input impedance, hum pick-up on the input lead would often be a problem when examining small signals in high impedance circuits unless a screened lead were used. However, one metre of screened lead could easily add another 50–100pF to the oscilloscope's input capacitance: even at a modest frequency, the reactance of this capacitance can be embarrassingly low. On the other hand, trying to connect the circuit under test directly to the input connector of the scope with negligible lead lengths is always tedious and often impossible. The usual solution to this problem is the passive divider probe, and this is the first accessory described in this chapter.

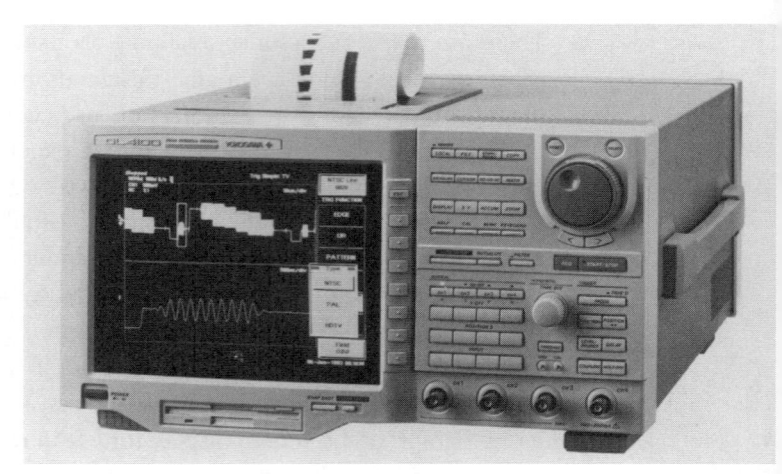

Figure 10.1
The four channel DL4100 features 10 bit sampling to 100Ms/s, providing a single shot bandwidth of 40MHz and a repetitive bandwidth of d.c. to 150MHz (300MHz for the DL4200, of similar appearance). A zoom facility makes good use of the exceptional vertical resolution, which can be increased to 11 bits with digital filter or 15 bits with averaging. Other standard features include 100kword per channel memory and both Zone and Parameter GO/NO-GO comparison functions, whilst options include a built-in printer and a SCSI interface. (Reproduced by courtesy of Martron Instruments Ltd.)

Accessories for use with digital storage oscilloscopes

Passive divider probes

Experience shows that to connect an oscilloscope to a circuit under test, a lead about one metre in length is usually convenient, screened to avoid hum pick-up when working on high-impedance circuits. Even a low-capacitance cable has a capacitance of about 60pF/metre, so a metre of cable plus the input capacitance of the scope would result in about 100pF of input capacitance all told. The purpose of a 10:1 passive divider probe is to reduce this effective input capacitance to around 10pF. This is a useful reduction, bearing in mind that at even a modest frequency like 10MHz the reactance of 100pF is as low as 160Ω.

Figure 10.2 (a) and (b) show the circuit diagram of the traditional type of oscilloscope probe, where C_o represents the oscilloscope's input capacitance, its input resistance being the standard value of 1MΩ. The capacitance of the screened lead plus the input capacitance of the oscilloscope form one section of a capacitive potential divider. The trimmer C_T forms the other, and it can be set so that the attenuation of this capacitive divider is 10:1 in volts which is the same attenuation as provided by Ra (9MΩ) and the 1MΩ input resistance of the oscilloscope. When this condition is fulfilled, the attenuation is independent of frequency, as shown in Figure 10.3 (a). Defining the cable plus scope input capacitance as C_E i.e. $C_E = (C_C + C_o)$, then C_T should have a reactance of nine times that of C_E i.e. $C_T = C_E/9$. If C_T is too small, high-frequency components (e.g. the edges of a squarewave) will be attenuated by more than 10:1. resulting in the waveform of Figure 10.3 (b). Conversely. if C_T is too large, the result is as in Figure 10.3 (c).

The input capacitance of the oscilloscope C_o is invariably arranged to be constant for all settings of the Y input attenuator. This means that C_T can be adjusted by applying a squarewave to the oscilloscope via the probe using any convenient Y sensitivity, and the setting will then hold for any other sensitivity setting. Many oscilloscopes provide a squarewave output on the front panel specifically for setting up passive divider probes. Such probes most commonly provide a division ratio of 10:1. but other values are sometimes found. e.g. the Tektronix P6009 100:1 probe operating to l20MHz with a maximum input capability of l.5kV, and the P6105A 75MHz 40kV probes. Some 10:1 probes have provision for shorting R_A and C_T to provide an alternative 1:1

Digital storage oscilloscopes

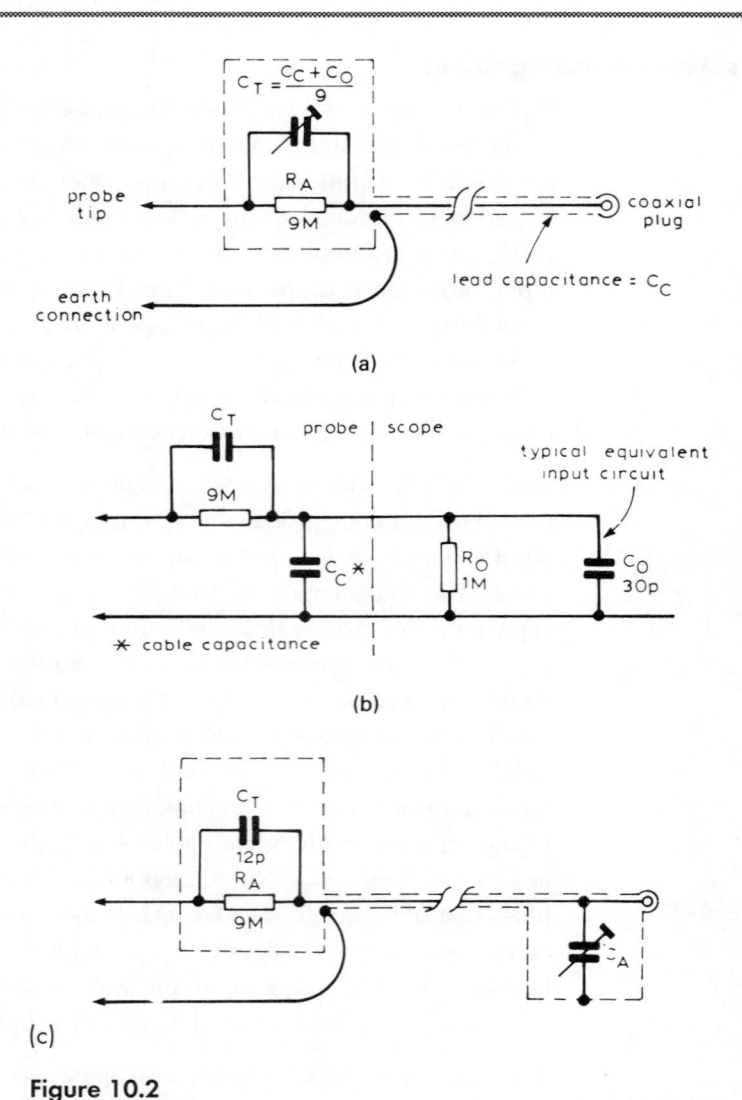

Figure 10.2
(a) Circuit diagram of traditional 10:1 divider probe. (b) Equivalent circuit of probe connected to oscilloscope. (c) Modified probe circuit with trimmer capacitor at scope end. (Reproduced by courtesy of *Practical Wireless*.)

ratio. When using such a probe in the 1:1 mode, the capacitive loading on the circuit under test is of course ten times as great as in the 10:1 mode, and its use is therefore confined mainly to lower frequencies.

The circuit of Figure 10.2 (a) provides the lowest capacitive circuit

Accessories for use with digital storage oscilloscopes

loading for a 10:1 divider probe, but has the disadvantage that 90 per cent of the input voltage (which could be very large) appears across the variable capacitor. Some probes therefore use the circuit of Figure 10.2 (c): C_T is now a fixed capacitor and a variable shunt capacitor C_A is fitted, which can be set to a higher or lower capacitance to compensate for scopes with a lower or higher input capacitance respectively. Now, only 10 per cent of the input voltage appears across the trimmer, which can also be conveniently located at the scope end of the probe lead, permitting a smaller, neater design of probe head.

Even if a 10:1 passive divider probe (often called a ×10 probe) is incorrectly set up, the rounding or pip on the edges of a very low-frequency squarewave, e.g. 50Hz, will not be very obvious, because with the necessary slow timebase speed the squarewave will appear to settle very rapidly to the positive and negative levels. Conversely, with a high frequency squarewave, say 1MHz, the division ratio will be determined solely by the ratio C_P/C_T. Waveforms as in Figure 10.3 will be seen at frequencies of around 1kHz. At very high frequencies, where the length of the probe lead is an appreciable fraction of a wavelength, reflections occur, since the cable is not terminated in its characteristic impedance. For this reason, oscilloscope probes often incorporate a resistor of a few tens of ohms in series with the inner conductor of the cable at one or both ends, or use a special cable with an inner made of resistance wire. Such measures are necessary in probes that are used with oscilloscopes having a bandwidth of 100MHz or more.

Special ×10 divider probes are available for use in pairs with an oscilloscope with a Y1-Y2 facility (Channel 1 plus Channel 2, with Channel 2 inverted). By effectively making both R_A and C_T adjustable (see Figure 10.1), the gain of the scope's two Y channels can be equalized at both high and low frequencies. For example the Tektronix differential probe pair P6135A with its 150MHz bandwidth can provide 20000:1 common rejection ratio (CMRR) from d.c. to 1kHz, derating to 100:1 at 20MHz, corresponding to 86dB and 40dB respectively. The P6046 active differential probe system, with its DC to 100MHz bandwidth and 60dB CMRR at 50MHz, converts from balanced floating signal to a single ended signal, referenced to ground. It thus needs only a single input channel on an associated oscilloscope. Figure 10.4

Digital storage oscilloscopes

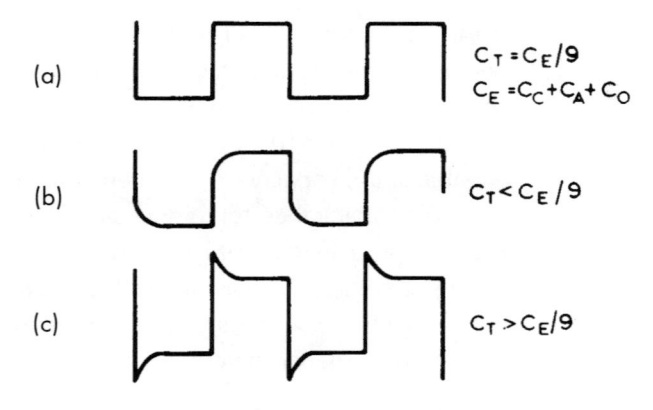

(a) $C_T = C_E/9$
$C_E = C_C + C_A + C_O$

(b) $C_T < C_E/9$

(c) $C_T > C_E/9$

Figure 10.3
Displayed waveforms with probe set up (a) correctly, (b) undercompensated, (c) overcompensated.

shows the probe head, together with the optional clip-on ×10 attenuator tip, which extends the input range to 10mV to 2V per division (in 1, 2, 5 steps) and the common mode linear dynamic range to ±50V.

While a 10:1 passive divider probe greatly reduces loading on the circuit under test compared with a similar length of screened cable, its effect at high frequencies is by no means negligible. Figure 10.5 shows the typical variation of input impedance versus frequency of such a 10:1 passive divider probe. Another point to watch out for when using such a probe is the effect of the inductance of its ground lead. This is typically 150nH (for a 15cm lead terminated in an 'alligator' clip), and forms a resonant circuit with the input capacitance of the probe. On fast edges, this will result in ringing in the region of 150MHz, so for high frequency applications it is essential to discard the ground lead and to earth the grounded nose-ring of the probe to circuit earth by the shortest possible route.

Not only the theory of operation of passive divider probes, but also their use has been covered here (rather than in Chapter 11) since they are by far the commonest — and to that extent the most important — oscilloscope accessory. Many a technician (and chartered engineer too) has wasted time wondering why the amplitude of a 10MHz clock waveform, for example, was out of specification, only to realize eventually that the ×10 probe being

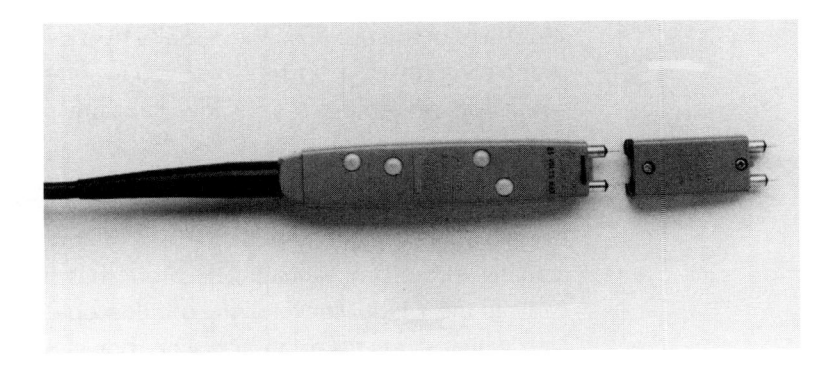

Accessories for use with digital storage oscilloscopes

Figure 10.4
The P6046 active differential probe system converts balanced floating signals to single-ended ground referenced signals. Shown here is the probe head, with the optional clip-on x10 attenuator. (Reproduced by courtesy of Tektronix Inc.)

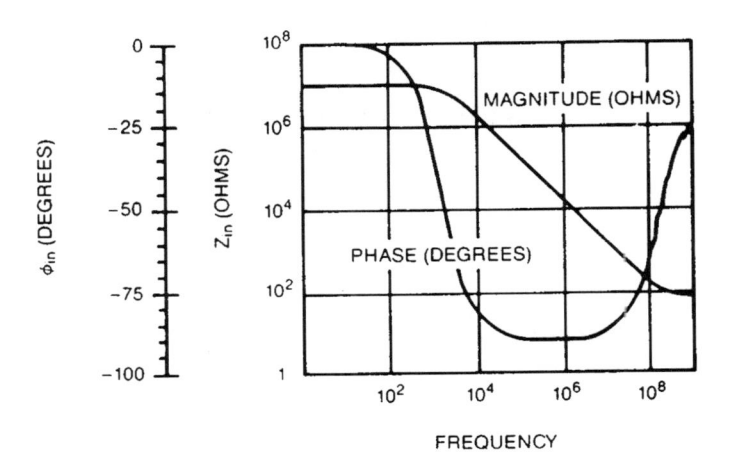

Figure 10.5
Variation of impedance at the tip of a typical X10 passive divider probe. (Reproduced by courtesy of Tektronix UK Ltd.)

used was not correctly set up for use with that particular oscilloscope!

Active probes

The reduced capacitive circuit-loading provided by the passive divider probe is dearly bought, the price being a reduction in the

109

sensitivity of the oscilloscope, usually by a factor of 10. An active probe can provide a 1:1 ratio, or even in some cases voltage gain, while still presenting a very low capacitive load to the circuit under test.

This is achieved by mounting a small unity gain buffer amplifier having a high input impedance and a low output impedance actually in the probe head. The probe has two leads, a coaxial cable to the Y input socket of the oscilloscope, and a power supply lead which connects to an accessory power socket, either on the oscilloscope itself or on a separate special free-standing probe power-supply unit. With the simple arrangement described, the maximum signal that can usefully be applied to the probe is obviously limited by the input voltage swing that the probe head amplifier can handle. This can usually be increased by the use of 10:1 or 100:1 divider caps, which can be clipped on to the probe's input. These not only increase the input voltage the probe can handle, but may also reduce the input capacitance even further.

The extensive Tektronix range of active probes includes types with bandwidths up to 4.0 GHz with an input capacitance of less than 0.4pF. The P6201 offers attenuations of 1:1, 10:1 and 100:1 all at 900MHz bandwidth and an input capacitance of 1.5 pF (3 pF at 1:1). The circuit of a typical active FET probe is shown in Figure 10.6.

Current probes

The probes described so far, both active and passive, are designed for the measurement of voltage waveforms. However, probes are also available which measure current waveforms, very useful, for example, if one is developing a switch-mode power supply. There are passive current probes, but these usually have low sensitivity and a limited frequency response that does not extend down to d.c., though they can be useful where these limitations are not important.

Current probes usually have a slotted head, the slot being closed by a sliding member, after slipping in the wire carrying the current to be measured. There is thus no need to break the circuit in order to thread the wire through the probe. Current probes produce an output voltage identical to the waveform of the current flowing in

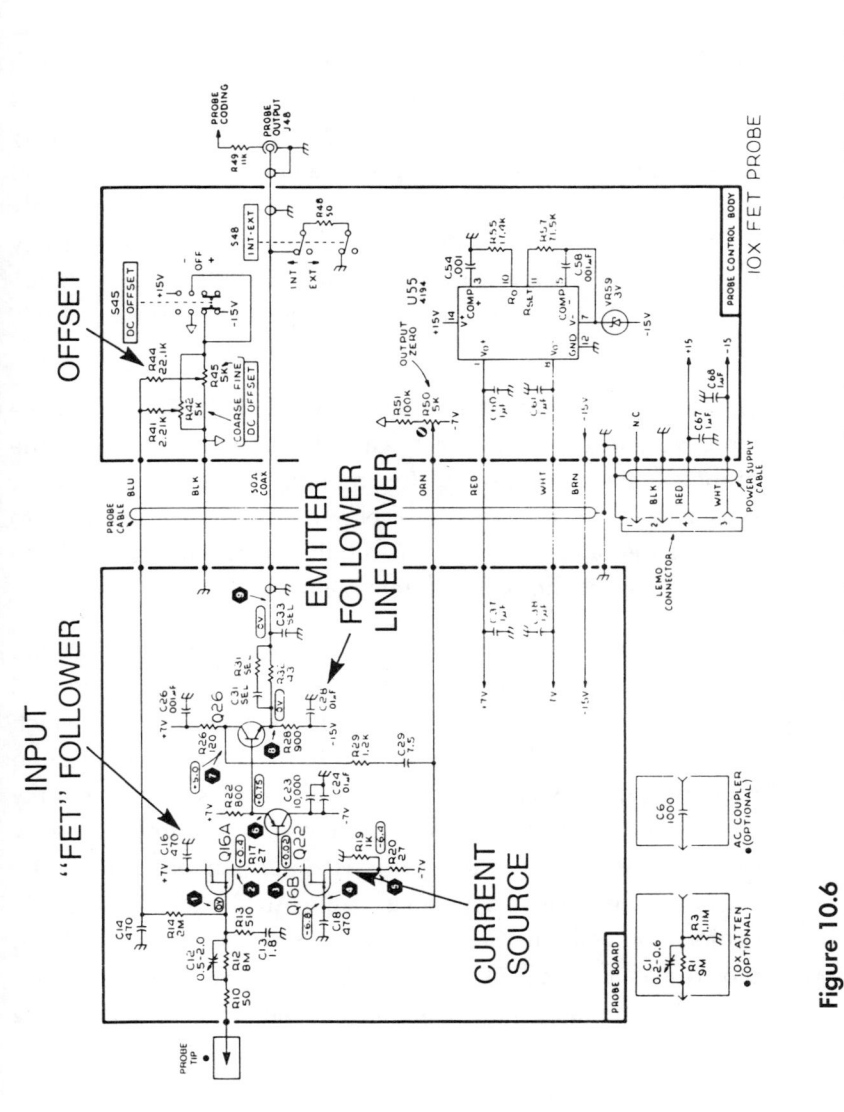

Accessories for use with digital storage oscilloscopes

Figure 10.6
The circuit of a typical active FET probe, the P6202A with a d.c.–500MHz bandwidth and a 2pf input capacitance. (Reproduced by courtesy of Tektronix UK Ltd.)

the wire. Some current probes work down to d c., others are a.c. only. A typical passive a.c. only probe can be plugged via its special passive termination directly into an oscilloscope, though in this mode the low-frequency cut-off point, depending on the particular probe, may be anywhere in the range from under 100Hz to 1kHz or more. For instance, the Tektronix P6021 has a bandwidth of 120Hz to 60MHz with a 5.8 nS risetime and offers sensitivities of 20mA/div and 100mA/div (with the scope sensitivity set to 10mV/div). However, special amplifiers are available to interface an a.c.-only probe to an oscilloscope; these not only increase the sensitivity of the probe, but extend its low-frequency cut-off point downwards by a factor of about l0. Thus the P6021 used with a suitable amplifier, has a frequency response of 12Hz to 38MHz, with the sensitivity increased to 1mA/div.

One type of current probe amplifier works by having a negative input resistance, which largely cancels out the resistance of the probe's sensing winding. By reducing the sensing-circuit resistance to near zero, a lower induced voltage suffices to produce the output, keeping the required magnetizing current to a negligible fraction of the current being measured.

Current probes with a frequency response down to d.c. are usually active types though most of the electronics is contained within an interface box to which the probe connects, and which has an output which can be applied to the Y input socket of an oscilloscope. As well as the usual split core as in an a.c.-only probe, there is a Hall element for the d.c. and low-frequency response, as indicated in Figure 10.7 (a). An example is the current probe system consisting of the A6303 probe which, in conjunction with the matching AM503B current probe amplifier, provides a bandwidth of d.c. to 50MHz and measures currents up to 20A continuous, 50A peak, Figure 10.8. The amplifier's 50Ω output will drive any oscilloscope, digital storage type or otherwise.

There is an important point to bear in mind when using current probes. When using ordinary voltage transformers, the volt-second product applied to the primary must be limited, to prevent core saturation. Thus a transformer designed for 440Hz use can only support one-ninth of its rated primary voltage if used on

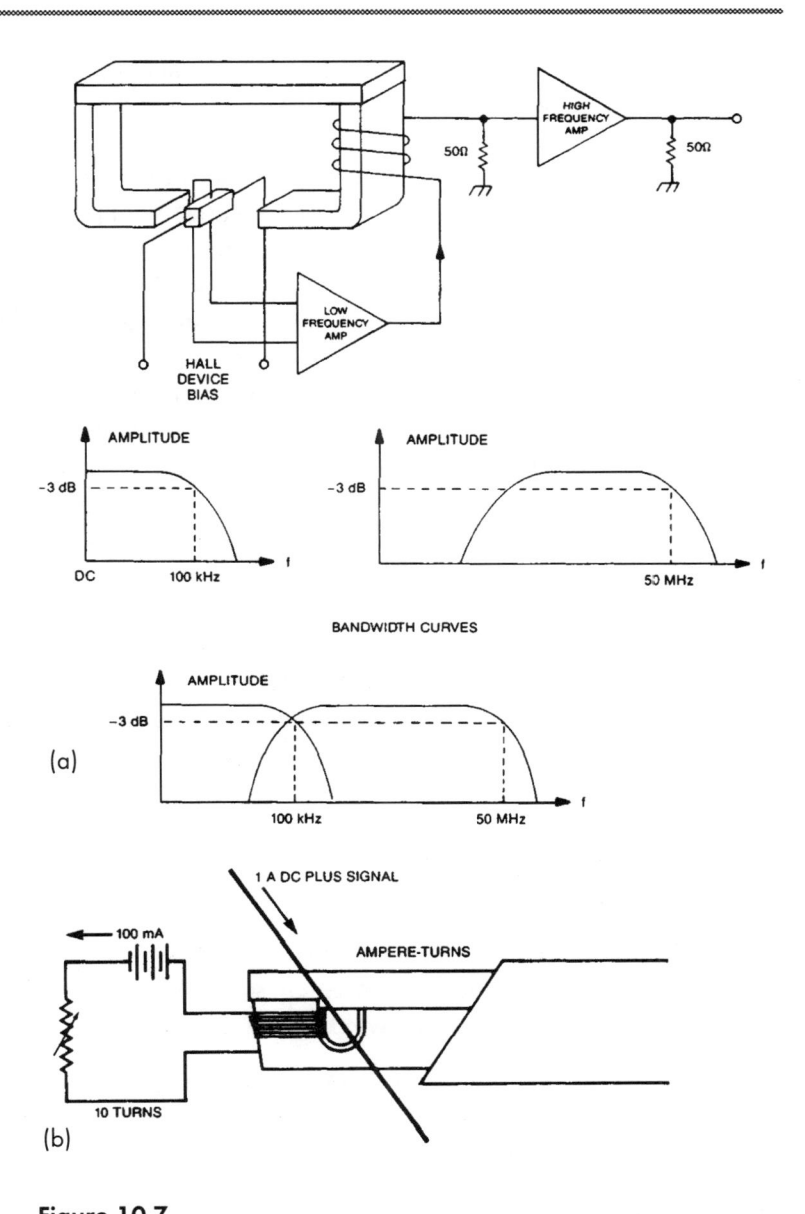

Accessories for use with digital storage oscilloscopes

Figure 10.7
(a) Current transformer and Hall-effect device combined to provide a wide bandwidth extending down to d.c. (b) Current bucking used to prevent core saturation is effective, but may affect h.f. performance. (Reproduced by courtesy of Tektronix UK Ltd.)

50Hz mains. For current transformers, including current probes, there is a corresponding Amp-s (amp-second) product, beyond which the core will saturate, and the output voltage will no longer be a true representation of the current waveform to be measured. For the passive a.c.-only probes mentioned, this limits the maximum current capability at low frequencies; for instance, the Amp-s product for the P6021 with passive termination is 500 x 10^{-6}. But this is extended to 0.5 Amp-s when using the current transformer CT-4 with the probe (with or without the amplifier). There is a limit of 100 x 10^{-6} Amp-sec also for the A6303 d.c. to 50MHz current probe used with the AM503B current probe amplifier with its 20A continuous and 50A peak current rating. However, a special feature of the AM503S Current Probe System utilizes the fact that the fluxes due to opposing currents are subtractive. The AM503S senses the current level in the conductor under test and feeds an equal but opposite current through the probe. This 'bucking' current nulls out the flux due to the current in the transformer and eliminates any core saturation. In the case of the A6303, the bucking current is effective up to a limit of 20 A. thus removing any concern for Amp-s product considerations regardless of frequency, except for currents over the 20A continuous rating up to the 50A peak rating.

Difficulty can arise when trying to measure an a.c. signal component riding on a larger d.c standing current. Current bucking can be used with any current probe to circumvent the problem, as indicated in Figure 10.7 (b), though the high-frequency response may suffer as a result, due to the presence of the additional winding.

Various current transformers are available to increase the current range that can be measured. For example, the CT4 extends the range of the P6021 to 20 000A peak.

Mains isolation

The Y input sockets on an oscilloscope normally have their outer screens connected to the instrument's metalwork and thus to the earth wire in the mains lead. Thus the input as it stands cannot be connected to circuitry which is at a different potential from mains earth, for example live-side components in a direct-off-line switchmode power supply. Hence the highly deprecated and very dangerous practice of disconnecting an oscilloscope's earth lead.

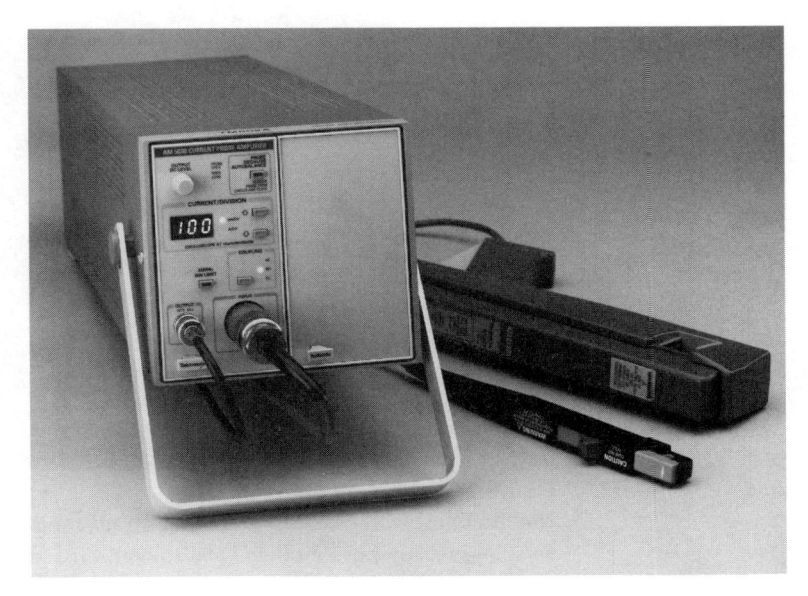

Accessories for use with digital storage oscilloscopes

Figure 10.8
The AM503S current probe system, comprising the AM503B current probe amplifier, together with probes A6302 (up to 50A peak pulse, <7ns risetime) and A6303 (500A, 23ns). (Reproduced by courtesy of Tektronix Inc.)

However, under specific conditions, safety standards do permit indirect grounding as an alternative to direct grounding. All of the grounding requirements apply, except that the grounding circuit need not be completed until the available voltage or current exceeds a prescribed amount. The Tektronix A6901 Ground Isolation Monitor fits between an oscilloscope and the mains, and continually monitors the voltage on the instrument's case/metalwork. The latter is permitted to float up to 40V peak, 28Vrms from ground. When this value is exceeded the mains supply to the instrument is interrupted, the isolated grounding system is connected to the supply grounding system, and an audible alarm is sounded. Applications include connecting the oscilloscope's grounding system to the -2V load-return reference rail instead of zero volts in ECL circuits, to reduce probe loading, and reducing hum problems in low level audio circuits by avoiding earth loops. This model is no longer current, though many are still in use. An alternative to mains isolation is *signal*

Digital storage oscilloscopes

Figure 10.9
The Gould 6600 Series ECG/Biotach signal conditioner is a versatile, multi-purpose module with isolated ECG and biological rate measurement capabilities.

isolation, and equipments for this purpose are described in Chapter 11.

A particularly critical application calling for mains isolation arises in medical electronics. In ordinary cases of electric shock, a current of just a few milliamps can be fatal. In the case of a patient with electrodes deliberately attached, possibly internal, there must be no possibility of any measurable mains-derived current passing through the electrodes and into the body, not only for reasons of patient safety, but also because the biophysical signals being monitored are extremely weak, and would easily be masked by hum. In this situation, special biomedical signal conditioners, such as that illustrated in Figure 10.9, are employed.

An alternative approach to mains isolation is to dispense with the mains lead entirely. This is possible with any digital storage oscilloscope which is capable of operating from internal batteries, such as those illustrated in Figures 10.10 and 10.11.

Normal safety rules should always be observed when working with high voltages, especially where the ground lead of the oscilloscope is at other than ground potential. It is unwise to work on equipment with voltages in excess of 50V if there is no other

Accessories for use with digital storage oscilloscopes

Figure 10.10
Described as a Graphical Multimeter, the economical Fluke model 863 incorporates a 32 000 count DMM, basic 1MHz digital oscilloscope, frequency counter and recorder. (Reproduced by courtesy of Fluke

Figure 10.11
The Fluke model 105 Scopemeter incorporates all the facilities of a wide range DMM, frequency meter and a two channel 100MHz digital oscilloscope. Facilities include autoset, cursor readout measurements, glitch detect, waveform storage, print-out and many others. (Reproduced by courtesy of Fluke Corporation.)

117

person in the same area. Yet another alternative is signal voltage isolation. This is covered in the next chapter.

Calibrators

A common accessory used with analog oscilloscopes is the oscilloscope calibrator. This is not frequently employed by the oscilloscope end-user, but is found in the test equipment lab of any large company and in the workshops of independent instrument repair and recalibration organisations. Essential for use with analog oscilloscopes (especially in the days when they used valves), calibrators are also used with digital storage oscilloscopes. However, the frequency with which they need be employed has been greatly reduced in many of the advanced models from the major manufacturers. These instruments incorporate elaborate self-checking and self-calibrating facilities, with an auto-calibration cycle being automatically initiated at switch-on, and following any substantial change in ambient temperature.

Operation of mains-powered instruments in mobile applications

Finally a word about using a mains-powered digital storage oscilloscope away from the lab, where a.c. mains supplies are not available. In these circumstances, there are two main options for obtaining a suitable supply. One is a portable generating set, powered by a small engine using either gasoline (petrol), or diesel fuel; the other is an inverter, powered typically by either 12V or 24V d.c. Operation from a generating set is usually no problem, but difficulty can arise when using an inverter. Some digital storage oscilloscopes use a switch mode power supply which operates without any voltage adjustment, from any input in the range 100–260Va.c., 50/60Hz, and these will generally operate satisfactorily from an inverter. But some digital storage oscilloscopes, especially older models, use a conventional mains-transformer plus rectifiers plus smoothing circuits, producing raw supplies fed to linear regulators for each of the internal voltage rails. The instrument is then designed to work from a specified user-set input voltage, such as 110, 120, 220 or 240Va.c., as available from the public electricity supply. The voltages quoted are rms, but the supply waveform is actually sinusoidal, giving — for example — a peak voltage of ±339V for 240V a.c. line voltage. On the other hand, many if not most inverters provide what is basically a square waveform, with just the corners rounded off a little. For a nominal 240Va.c. r.m.s. output, the peak output voltage from such an inverter will be little more than ±240V. Thus the raw supplies may be inadequate for proper functioning of the regulators, and the stabilised supply rails thus too low and badly corrupted with hum. This will prevent proper operation, and may well cause the microcontroller to crash, preventing the instrument operating at all. Attempting to solve the problem by switching the mains input to a lower setting, such as running the instrument from a nominal 240Va.c. squarewave inverter with the instrument's input selector set to 220V is not advisable. To do so exposes the transformer primary to an excessive volt.second product, risking core saturation and possible permanent damage to the instrument. The cure is either to use a generator, or switch to an inverter supplying a sinewave output. Unfortunately, sinewave inverters are both more expensive and less efficient than the usual variety.

Chapter 11

Using digital storage oscilloscopes

This chapter considers various important aspects of using digital storage oscilloscopes, and notes how different models include different facilities. This will affect not only how you use them, but even how you choose which model to acquire, when considering purchasing or leasing. Most models are very versatile, and to that extent may be considered as general purpose instruments, while others — some of which are illustrated in these pages — are primarily dedicated to a particular use, such as those digital storage oscilloscopes specifically designed for working with the high voltages encountered in power supplies and mains distribution systems.

Early digital storage oscilloscopes were not only limited in performance, but operating them could be a little tricky. Modern instruments are much improved in this respect, but there are still points that the user should be aware of and watch out for. One of these is the possibility of aliasing, which can occur not only when the instrument is operating in equivalent time mode, but also in realtime mode, if the Nyquist criterion is violated, i.e. if the input signal contains frequency components higher than half the sampling rate.

This is graphically illustrated in Figure 11.1, where (a) shows a display of $2^1/_2$ cycles of (apparently) a 50Hz sinewave, displayed at 5ms/division. At this timebase speed, the oscilloscope is sampling at 10ksamples/sec, corresponding to 512 displayed

Using digital storage oscilloscopes

points across the screen. But despite the convincing looking display, the frequency of the input sinewave is actually nearer 13MHz (actually 12.9904MHz).

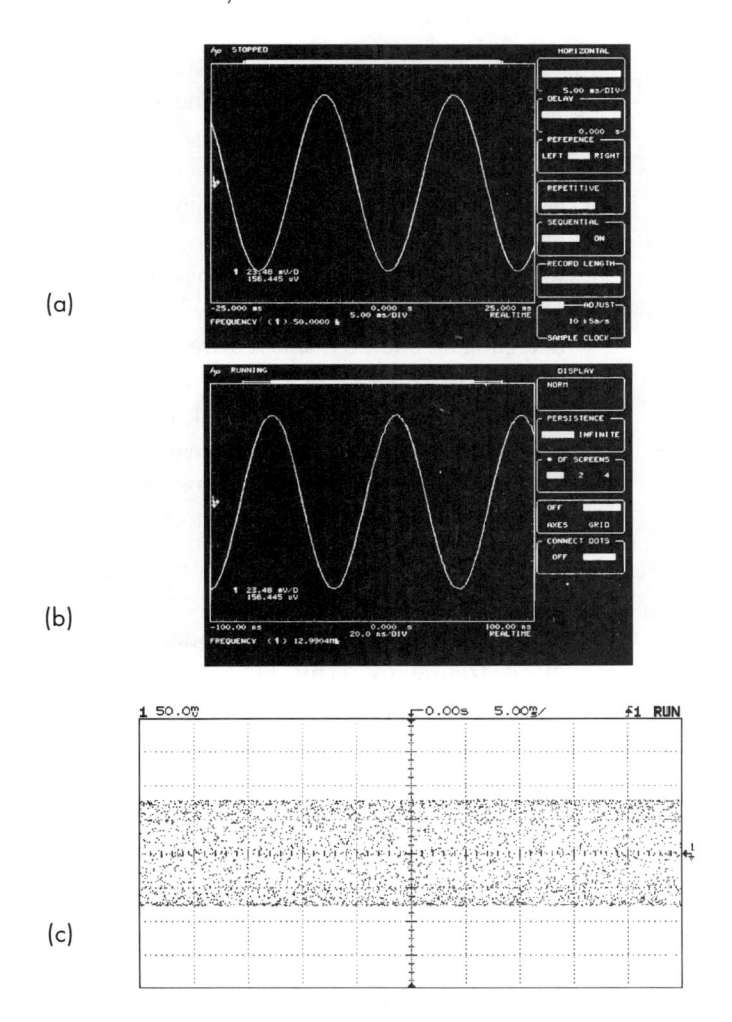

(a)

(b)

(c)

Figure 11.1
(a) With a timebase speed setting of 5ms/division, this looks like a display of a 50Hz sinewave. (b) The same waveform displayed at 20ns/division reveals it in fact to have a frequency of around 13MHz. (c) With the same input signal and timebase setting as (a), an aliased display does not appear on the HP54600 digital storage oscilloscope, thanks to its proprietary anti-alias algorithm. (Reproduced by courtesy of the Hewlett-Packard Company.)

121

Digital storage oscilloscopes

Given the 10ks/s sample rate, this grossly violates the Nyquist criterion, resulting in the misleading aliased display shown in Figure 11.1 (a). The cause is the fact that the nearest even harmonic of the sampling rate to the input signal differs in frequency from it by 50Hz.

Note that in Figure 11.1 (a) the timebase is stopped, i.e. the oscilloscope is not running in repetitive mode. But even if it were, a stable, convincing display could still result if the above relationship between the sampling and signal frequencies were exact. If it were very nearly, but not quite so, the aliased waveform would be seen drifting across the screen, as though the oscilloscope were not properly triggered. This is one clue to detecting an aliased display: always check for a solid stable trigger. Another is to check that the waveform is unchanged (except for being stretched in the horizontal direction) as the timebase speed is increased. Thus, for example, with the timebase speed increased to 20ns/div (at which setting the oscilloscope is sampling at 2Gs/s), the waveform is displayed in its true 13MHz form, Figure 11.1 (b).

In the above example of an aliased display, the input frequency was close to a harmonic of the sampling frequency, say the nth harmonic, i.e. the sampling frequency was around one nth of the input frequency. The Hewlett-Packard oscilloscope model HP54600 incorporates a patented proprietary anti-alias algorithm, so that under the same conditions as in Figure 11.1 (a), an aliased image does not appear, see Figure 11.1 (c).

Another way of detecting an alias is to set the oscilloscope to acquire the signal in max/min mode (available on many models). In this mode, the sample rate is the maximum of which the instrument is capable, regardless of the actual sweep speed setting. At each even (or odd) storage location, the oscilloscope stores in memory the maximum (or minimum) value detected during that sample interval. The resultant acquisition is displayed in dot-joining mode. Since (in the above example, for instance), during any one sample period there are several or many complete cycles of the input waveform, the positive peak value is held in alternate locations, and the negative peak value in the others. Thus the oscilloscope displays a sawtooth waveform with one cycle per two storage locations. With 50 or more such points per horizontal division, the display simply looks like a band of light, of height equal to the peak to peak value of the input waveform; no aliased waveform can appear.

Using digital storage oscilloscopes

Some digital storage oscilloscopes have a feature which provides a warning of the possibility of an aliased display. In addition to being fed to the Y input amplifier and digitising section of the instrument, the input signal is also applied to a frequency counter section. This compares the measured frequency of the input with the sampling rate which applies at the selected timebase speed. In the event that the sampling rate is inadequate, a screen read-out warning appears. This scheme works well with simple repetitive waveforms such as sinewaves and squarewaves, but can be fooled by complex waveforms with a great deal of waveshape detail in each basic cycle of the waveform. In other digital storage oscilloscopes, an *autoset* button is provided. This causes the instrument to measure the peak to peak voltage of the input waveform, and its frequency, and select appropriate values of Y deflection factor and timebase speed automatically. This generally works well, except, again, with certain difficult waveforms.

Some oscilloscopes offer an absolutely foolproof way of checking that a given waveform display is real, and not an alias. These are the dual mode oscilloscopes, which are basically an analog oscilloscope with built in digital storage adaptor. If there is any doubt about the veracity of the display in digital storage mode, the user can switch immediately to the real-time analog mode with the same Y sensitivity and timebase speed settings — then switch back to digital storage mode to enjoy the peculiar virtues thereof, such as pre-trigger view and waveform averaging etc. The combined digital and analog dual mode oscilloscopes produced by Fluke (of which some are illustrated) go under the trade-name *Combiscope*, while one or two other manufacturers also produce dual mode oscilloscopes.

The above schemes all aim at avoiding the user being misled by an aliased display when the Nyquist criterion is violated. However, an alternative approach is to try and ensure that it is not violated, namely that there is negligible signal energy at frequencies higher than half the sample rate. This can be brought about by low-pass filtering the signal before it is sampled and digitised. Ideally, the filter should have a constant group delay to avoid introducing ringing on fast edges. A Gaussian filter fits the requirement, and if the -3dB point is set at one quarter of the sampling frequency, very little energy at frequencies above the Nyquist rate, at which the response would be -12dB, will be passed, see Figure 11.2. Very few digital storage oscilloscopes

Digital storage oscilloscopes

include a front-end filter which can be set to any desired cut-off frequency, but the choice of a few cut-off frequencies is often available. For example, the Hewlett-Packard 100MHz digital storage oscilloscope type HP54600A has a user selectable 20MHz bandwidth limit, and the bandwidth of the Tektronix 1000MHz model type TDS684B can be restricted to 250MHz or 20MHz at will. At its full 1000MHz bandwidth, it samples at 5Gs/s, so even at the fastest timebase speeds, the roll-off of the front-end frequency response provides some low-pass anti-aliasing filter action.

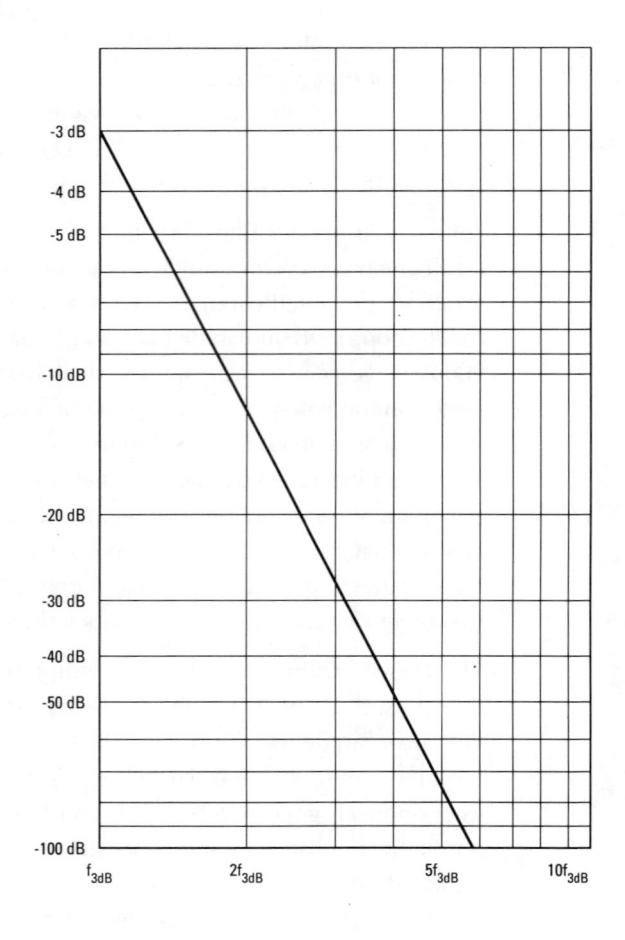

Figure 11.2
Magnitude response of a Gaussian filter in the frequency domain. (Reproduced by courtesy of the Hewlett-Packard Company.)

Triggering is such an important aspect of using a digital storage oscilloscope that, although it has been covered in some detail in a previous chapter, some further comments here will not be out of place. The commonest form of triggering used is edge triggering, where the oscilloscope triggers, on each sweep, on the first edge that it sees crossing the user-set trigger level, in the selected direction (e.g. positive-going). This is fine for a simple repetitive waveform such as a sine or squarewave. But with a more complicated sequence of events, it can lead to a jumbled, confusing display, Figure 11.3 (a). Using *hold-off*, the oscilloscope is not available to be retriggered immediately following the completion of the sweep, but only after the expiry of the further hold-off time. When the latter is suitably adjusted, the displayed traces are all triggered on the same edge of a complex waveform, as shown in Figure 11.3 (b).

Likewise, glitch capture has been dealt with earlier, but again is such an important subject that there is more to be said about it. The point to bear in mind is that, at the higher timebase speeds, there is a marked difference in performance between run-of-the-mill digital storage oscilloscopes and their analog counterparts. An analog oscilloscope, once triggered, ignores further triggers during the sweep time and also the following retrace or flyback time. It is only available to be retriggered, following any timebase sweep, once the retrace is complete. The retrace generally occupies less time than the sweep itself, typically 20% of the time, rising to nearer 100% at the very fastest sweep speeds. In order to obtain a clear, stable display of a complex waveform, the time before the next trigger is recognised and the next sweep can begin may be extended beyond the end-of-retrace, by means of hold-off, as described above. Nevertheless, in the absence of hold-off, an analog oscilloscope will see the next valid trigger shortly after the completion of the current sweep. So it is 'looking at' the input signal most of the time. This applies to a digital storage oscilloscope also, at least at slow sweep speeds. But a typical digital storage oscilloscope has to break off after acquiring a sweep, to transfer the data from the acquisition memory to the display memory (which, for economy, is usually much slower than acquisition memory), quite possibly to perform some digital signal processing upon it, and to format it for display. These chores are time-consuming, with the result that a run-of-the-mill

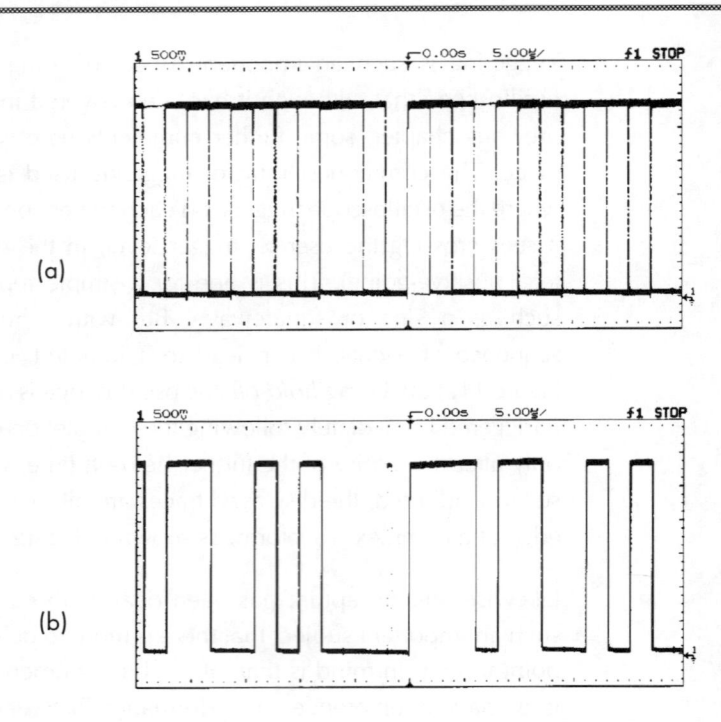

Figure 11.3
(a) A digital waveform may be repetitive over a period containing several pulses of different widths. Simple edge triggering then produces a confusing, overlapped display. (b) delaying the trigger by an appropriate amount after completion of the sweep, using the HOLD-OFF control, produces the desired display. (Reproduced by courtesy of the Hewlett-Packard Company.)

digital storage oscilloscope will typically make a maximum of somewhere between fifty to a few hundred acquisitions per second. This means that, except for quite slow sweep speeds, the instrument is only actually acquiring data for a tiny fraction of the total time. As a result, however sophisticated the trigger arrangements, it is unlikely to catch a rare glitch, because most of the time it is not looking at the input, as illustrated in Figure 11.4. Figure 11.5 compares the number of acquisitions per second of the Tektronix TDS700 series instruments with that of conventional digital storage oscilloscopes and with the fastest analog oscilloscopes. A special architecture, using proprietary ASICs, reduces dead time between acquisitions to 1.7μs, allowing these instruments to acquire and display more than 400 000

Using digital storage oscilloscopes

Ta = Time to acquire the data

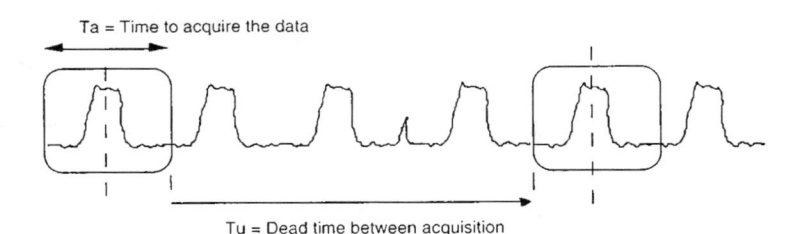

Tu = Dead time between acquisition

Figure 11.4
Illustrating how the dead time between sweeps can make capturing an infrequent glitch a very hit-and-miss affair. In a real life situation, Tu could be hundreds or even thousands of times Ta.

acquisitions per second. Figure 11.6 shows infrequent runt pulses captured and displayed on a TDS744A, where on an earlier model TDS544A, with its less frequent acquisitions, they are not captured, or at best would only be so after a very long period waiting for one to coincide with an acquisition sweep.

A point to bear in mind when using a digital storage oscilloscope — and still more when choosing one- is the effect of equivalent time operation when viewing logic signals, for example the D, Q and clock signals of a type D flip-flop, or the chip select to data valid delay in a memory system.

An oscilloscope which uses equivalent time sampling to obtain its high frequency performance, will build up a picture of the waveforms under investigation over a number of sweeps, triggered by, say, the clock waveform of the D flip-flop. The clock waveform is a truly repetitive waveform and so will be reproduced accurately. But the Q output will vary from acquisition to acquisition, depending upon the D input on each particular clock pulse.

An earlier chapter described how, in equivalent time sampling, points are acquired on each acquisition, until the complete picture is built up. Realtime and two forms of equivalent time sampling are illustrated in Figure 11.7. The top trace illustrates realtime sampling — here the sample rate of the oscilloscope is sufficient to acquire all the samples needed to define the waveform on a single pass, a mode of operation permitting single shot acquisition of a unique, non-recurring event. The middle

Digital storage oscilloscopes

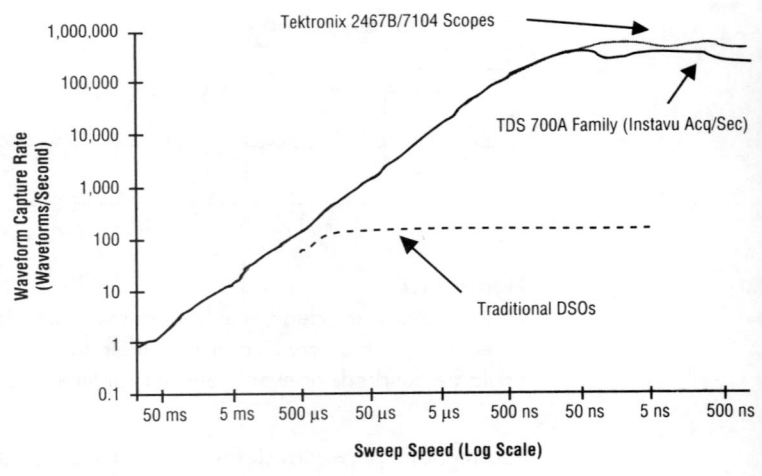

Figure 11.5
Comparing the number of waveform acquisitions per second versus sweep speed for a run-of-the-mill digital storage oscilloscope, a top-of-the-range analog oscilloscope, and models of the TDS700A family. The latter incorporate a 'InstaVu ‰', a special architecture which reduces dead time between sweeps to 1.7ms. (Reproduced by courtesy of Tektronix Inc.)

trace shows sequential sampling, where a point is acquired on one trace and the next point acquired (with a slightly greater delay relative to the trigger point) on the next acquisition and so on. This is the type of operation found in digital sampling oscilloscopes, where the next acquisition may occur not on the very next trigger event detected, but *n* triggers later, where *n* could be a very large number. The lower trace shows random repetitive sampling (also known as *random interleaved sampling* — RIS), where a subset of points is acquired on each pass. Just three passes or acquisition sweeps are shown, but further subsets would be acquired, building up to the same dot (sample) density as in the other traces pictured in Figure 11.7.

The effect of this equivalent time mode is clearly evident in Figure 11.8 (a), where following a chip select pulse, the chip mostly outputs a logic 1, but occasionally a logic zero. The result is a ragged display showing confusing artefacts. It is clear from the lower trace that this equivalent time oscilloscope acquires about twelve points per acquisition, at the particular timebase setting in use. By contrast, Figure 11.8 (b) shows the same data acquired on

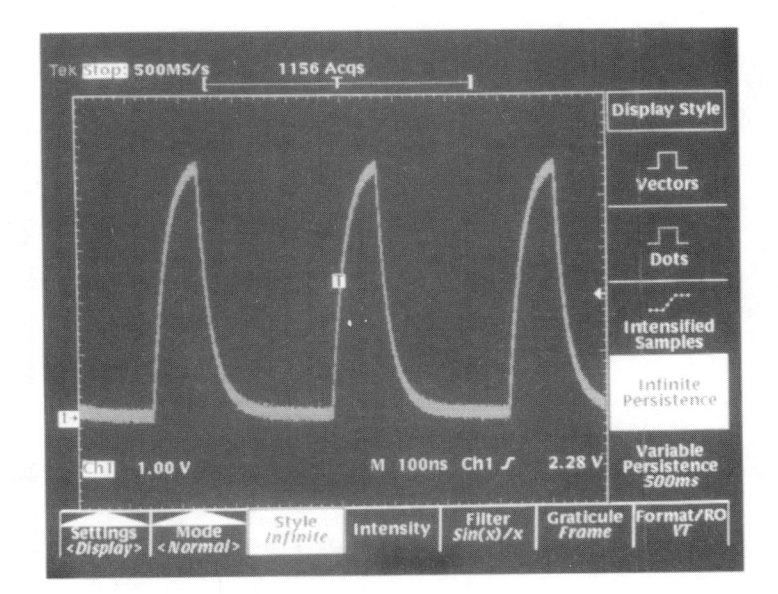

Using digital storage oscilloscopes

(a)

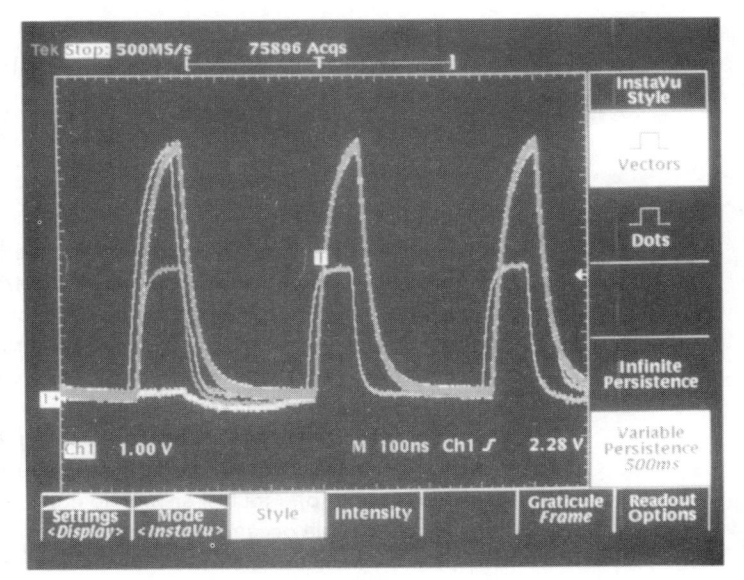

(b)

Figure 11.6
(a) A 500MHz bandwidth TDS544A oscilloscope does not show a very infrequent glitch, even after a considerable period in infinite persistence mode. (b) The same glitch is rapidly captured and displayed, even in variable persistence mode, by model TDS744A. (Reproduced by courtesy of Tektronix Inc.)

a digital storage oscilloscope with a high enough sampling rate to permit the waveform to be acquired in its entirety on a single pass,

Digital storage oscilloscopes

(a) Real-Time Sampling

(b) Sequential Sampling

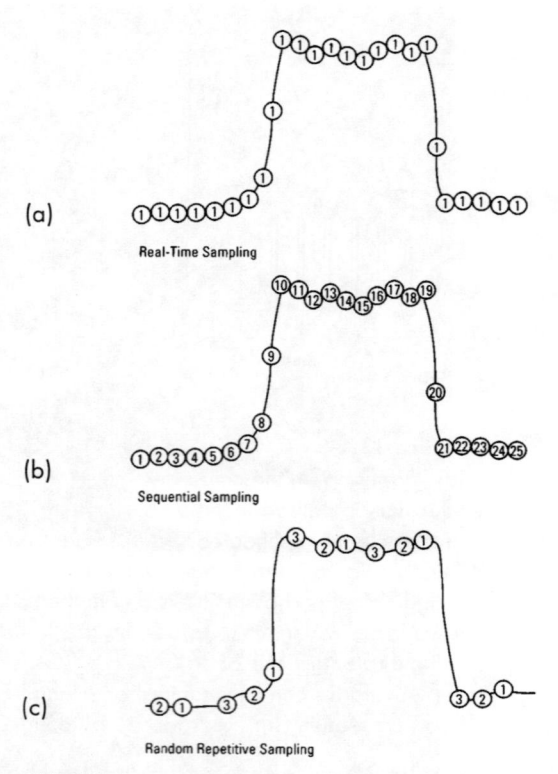

(c) Random Repetitive Sampling

Figure 11.7
(a) In realtime sampling, all the samples required to delineate a waveform are acquired in a single sweep, covering a single occurrence of the waveform. Provides single shot capture of a unique event. (b) In sequential sampling, points are captured (with successively greater delay relative to the trigger event) on successive occurrences of the waveform, or on every nth recurrence of the waveform. Digital sampling oscilloscopes use sequential sampling. Like (c) below, it only works with repetitive waveforms. (c) Random repetitive sampling falls between (a) and (b). Not fast enough to capture all the required samples on a single occurrence of the waveform, it can nonetheless acquire more than just one. Thus a delineation of the complete waveform can be acquired from far fewer recurrences of the waveform under examination than would be the case with sequential sampling. Digital storage oscilloscopes use random repetitive sampling to provide a higher bandwidth than would be possible in single shot mode. (Reproduced by courtesy of the Hewlett-Packard Company.)

i.e. using realtime sampling which is capable of capturing the waveform on a single shot basis.

An important consideration for the user of a digital storage

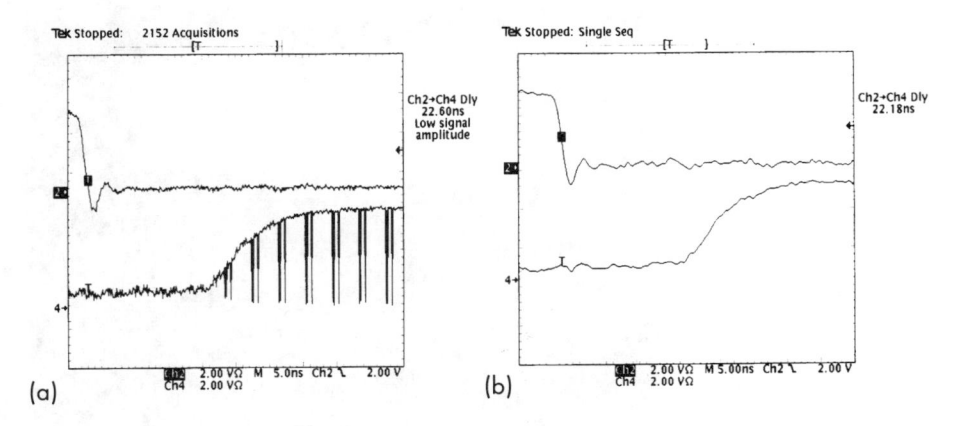

Using digital storage oscilloscopes

Figure 11.8
(a) Triggering on the chip select pulse, the chip select waveform is effectively repetitive (even if not recurring a precisely repetitive rate), and so cleanly delineated. The associated data output from the chip is usually a logic 1 (true), but occasionally a zero (false) resulting in the ragged display shown. This is due to the oscilloscope acquiring the trace in equivalent time. (b) On an oscilloscope with a high enough sample rate (e.g. the TDS640 with a 2Gs/sec. maximum sample rate) the waveforms can be acquired on a single shot basis, avoiding the artefacts seen in (a). (Reproduced by courtesy of Tektronix Inc.)

oscilloscope is the record length, i.e. the depth of memory, that it provides for each channel. This becomes of crucial importance if the application requires a large degree of horizontal expansion. A digital storage oscilloscope normally displays, at any timebase speed setting, the full acquired record. Thus the number of points in the record, in conjunction with the selected timebase speed, determines the sample rate — and hence the usable bandwidth, determined by the Nyquist criterion. Some digital storage oscilloscopes have a record length as short as 1K (1024 points) or even 512 points, while others have 16K, 50K, 250K or more, up to several Mbytes in some models. Figure 11.9 (a) shows a complete TV field captured on an instrument with 2Mbytes of acquisition memory (upper trace) with part of the record expanded horizontally to show a single line. Note that even the colour burst is captured. By comparison, Figure 11.9 (b), the same test on an instrument with only 50KBytes of acquisition memory clearly demonstrates the effect of undersampling.

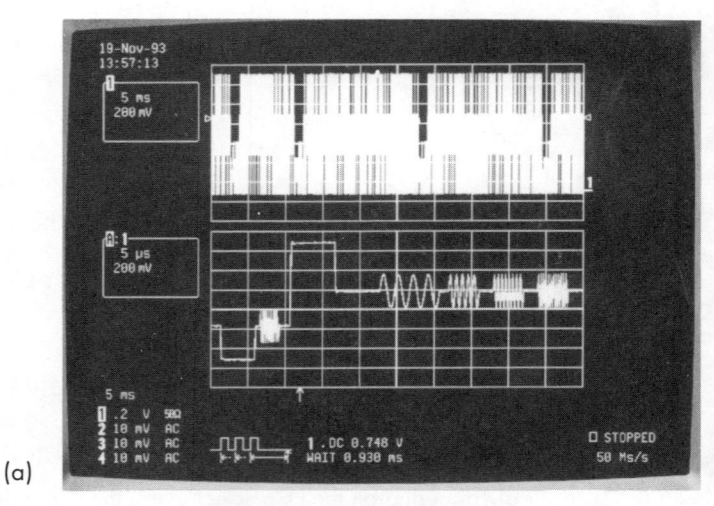

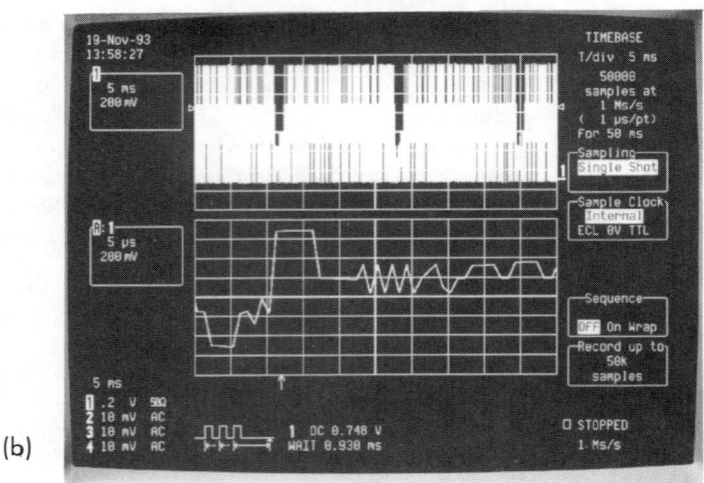

Digital storage oscilloscopes

Figure 11.9
(a) On a digital storage oscilloscope with a deep 2MByte memory such as the LeCroy 9354L, a complete TV field can be captured (upper trace) and expanded (lower trace) to the point where even the colour burst is visible. (b) On the less expensive 9350, with its 50kByte memory, the same test shows clearly the effect of undersampling. Like (a), the illustration shows a single shot acquisition. (Reproduced by courtesy of LeCroy Corporation.)

Such very long acquisition memories are unusual, but many manufacturers are moving away from the shorter lengths, such as 1K, with 16K now being commonly provided. This still means that on the slower timebase speeds, the sample rate is way below the

Using digital storage oscilloscopes

maximum of which the instrument is capable. The consequences of this, in terms of missed glitches or aliasing, and the use of max/min mode to combat them, have already been mentioned. This mode runs the acquisition system at its maximum rate at all times. But only one value (the highest or lowest encountered during sample periods, alternately) is recorded per acquisition memory location, i.e. per point displayed on the screen. The remainder of the samples taken (the vast majority) are simply discarded.

But there is another mode of operation where use is made of all the samples taken, even at the slowest sweep speeds. Instead of storing just one sample during the sample interval and simply discarding the rest (a process called *decimation*), all the sample values taken during the sample interval may be summed, and the result divided by the number of values n. This scheme is in incorporated in certain Tektronix models, the process being carried out in dedicated 'HI RES' hardware ahead of the acquisition memory, Figure 11.10 (a). This provides decimation with averaging, reducing the high speed stream of 8 bit data from the A/D converter to a slower stream of 16 bit data. The result, stored in each location of the acquisition memory, is the average value during the sample interval, which ideally, in a noise free system, should be the same as the sample taken halfway through the sample interval. When noise is present, decimation with the samples averaged over the decimation interval provides improved performance, relative to simple decimation, see Figure 11.10 (b). Averaging not only reduces the effect of the noise inevitably present in the system, it provides additional bits of resolution. This mode may thus may be used both to reduce the noise accompanying the wanted signal (assuming that measurement of the noise statistics is not required), and — where the signal is not too noisy to start with — to provide increased resolution. If the noise originally present is white, the degree of noise reduction will correspond to the increase in resolution. If the noise energy is concentrated at frequencies well above the Nyquist rate, the improvement will be greater than this, whereas if the noise is all well below the Nyquist rate, the improvement will be less.

Figure 11.11 shows how even a modest degree of *averaging* (or more accurately, smoothing) can make a marked improvement in

133

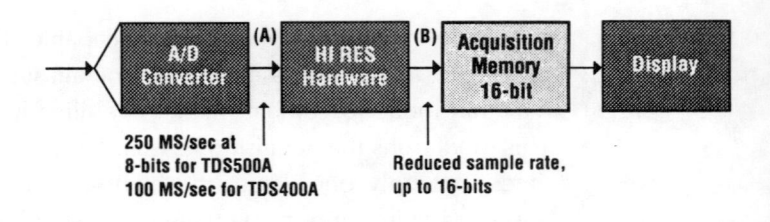

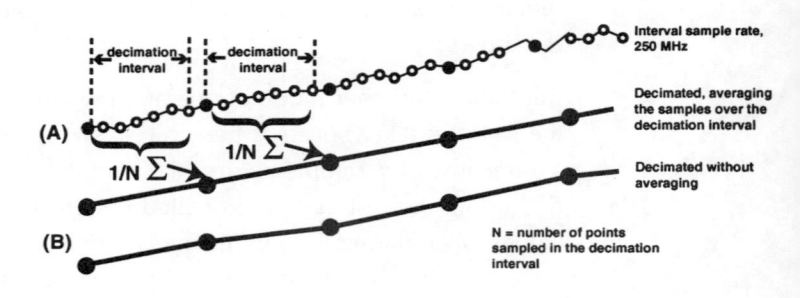

Digital storage oscilloscopes

Figure 11.10
(a) At slower sweep rates, the sample rate is normally well below the maximum of which instrument is capable. In these circumstances, both noise reduction and increased vertical resolution can be obtained, by sampling at the maximum rate and processing the data in dedicated 'HiRes' hardware before storing in acquisition memory. (b) Decimation with averaging over the sample interval provides increased resolution and noise reduction. This is achieved by sampling at the maximum rate, rather than the much lower rate which would suffice for one sample per sample interval. (Reproduced by courtesy of Tektronix Inc.)

the fidelity of waveform reproduction. The upper trace is of a positive-going step exhibiting about 12% overshoot, with ringing. The two lower traces show the tip of the edge with ×5 vertical expansion and ×25 horizontal expansion. The middle trace is as recorded, without resolution enhancement. The lower trace shows the effect of enhancement to one extra bit of resolution, as implemented in the LeCroy 9420 series oscilloscopes. The resolution enhancement is implemented by a set of linear phase FIR filters, the degree of resolution enhancement applied being user selectable from zero to 3 bits, in half bit increments. The one bit enhancement illustrated in Figure 11.11 is implemented with an FIR filter length of 5 samples, and reduces the -3dB bandwidth to 0.241 times the Nyquist bandwidth. The FIR filter

Using digital storage oscilloscopes

implementation is more efficient in terms of bandwidth than a simple moving average type of filter. The reduction in bandwidth will be of no significance if the wanted signal contains no components outside the filter bandwidth. If the signal of interest is strictly repetitive, e.g. a pure sinewave or jitter-free squarewave, it can usefully be acquired in equivalent time mode, using random interleaved sampling (RIS). By making the sample interval Δt very short compared to the period of the waveform of interest, the signal will be heavily oversampled. An FIR filter with, for example, a length of 117 samples can then be employed, giving a 3 bit increase in resolution. The bandwidth will be reduced to 0.016 times the Nyquist rate, but due to the large degree of oversampling, the wanted signal still lies wholly in the passband. Noise reduction has thus been achieved without loss of effective bandwidth.

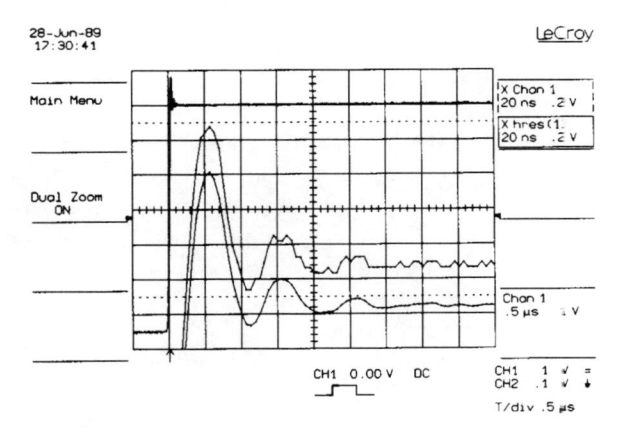

Figure 11.11
The top trace shows a relatively noise-free step response, the two lower traces the same but with x5 vertical expansion, x25 horizontal. The middle trace is an unprocessed expansion, the lower has 1 bit of enhanced resolution, showing the advantage of even a modest amount of resolution enhancement. (Reproduced by courtesy of LeCroy Corporation.)

It is sometimes necessary to take measurements of relatively small voltages which are not referenced to earth. An example is verifying the correct gate drive to SCRs, GTOs or IGBTs in high power electrical installations, where the voltage waveform to be monitored is at a very high potential, either d.c., a.c. or both,

Digital storage oscilloscopes

relative to ground. In such cases, the circuit to be monitored cannot, for safety reasons, be connected directly to an oscilloscope. Various manufacturers market *voltage isolators* or *ground isolators* to circumvent the problem. The Tektronix A6902B Voltage Isolator has been replaced by the A6907 (four input channels, see Figure 11.12) and A6909 (two channels) Voltage Isolators. These systems use a combination of transformer- and opto-coupling to provide a response from DC to 60MHz. Designed for use with any two or four channel oscilloscope, they permit simultaneous observation of signals at potentials of up to 850V (DC + peak AC), providing 60dB of common mode rejection ratio at 1MHz and 50dB at 10MHz. Two channels can also be combined to function as the input to a differential amplifier or an oscilloscope with a differential input, for floating differential measurements. Where greater separation between the circuit under test and the oscilloscope is necessary, the A6905S single channel fibre optic isolator can be used — up to 100m — or the A6906S, up to 200m.

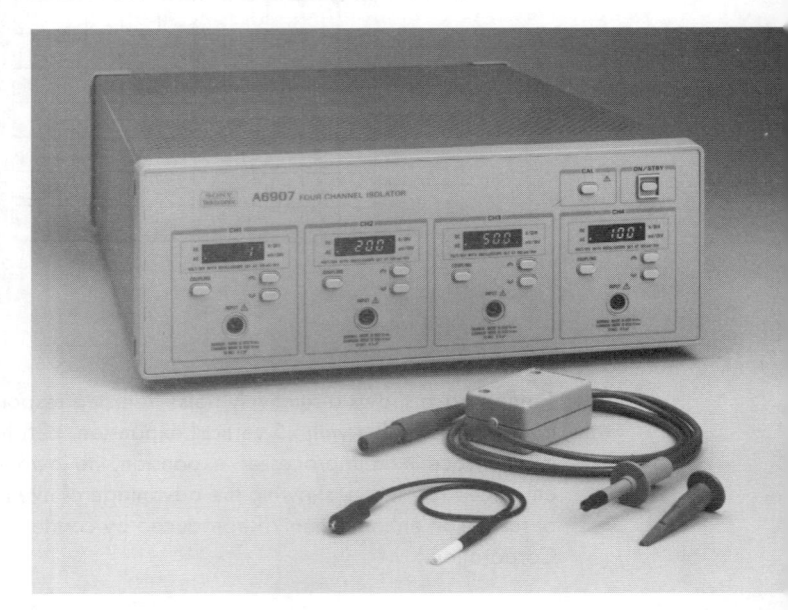

Figure 11.12
The A6907 Four Channel Isolator accepts, on each channel, differential voltages up to 850V with common mode voltages up to ±850V, producing ground-referenced outputs suitable for application to any oscilloscope. (Reproduced by courtesy of Tektronix Inc.)

Using digital storage oscilloscopes

Figure 11.13
The battery operated fibre-optic isolated ISOBE3000 Isolator provides nearly unlimited common mode voltage rejection, producing ground-referenced outputs suitable for application to any oscilloscope. Providing up to 50m remoting, it is available either in the single channel version transmitter/receiver pair illustrated, or in a rackmount version with up to eight channels.(Reproduced by courtesy of Nicolet Technologies Ltd.)

Another ground isolator (Figure 11.13) is the ISOBE3000 from Nicolet, designed for various applications, including taking measurements in hazardous environments. This fibre optic isolated probe system, offering nearly unlimited common mode voltage rejection, can be used where high voltages, high EMI or high explosives are involved. The makers list motor controls, power distribution, ordnance and even physiology among its applications, but note that for measurements involving human physiology, stringent safety rules apply. An alternative to a ground isolator used in conjunction with any normal oscilloscope, is provided by an oscilloscope specially designed to work with high voltages. These are produced by a number of manufacturers, including Nicolet, Siemens and Dranetz. For example, the Nicolet Power Pro 610 accepts up to four plug-ins, each having a differential input with >60dB common mode rejection ratio up to 10kHz and accepting a common mode input up to 10× full scale at any sensitivity. Sensitivity ranges from 10mV to 5V/division (with ×1 probe), up to 500V/division with ×100 probe. Safe overload on any input, with ×1 probe, is ±500V peak.

Glossary

Accuracy of an ADC: the difference between the analog input theoretically required to produce a given digital output and the analog input actually required to produce that output.

Accuracy of a DAC: the difference between the analog output that is expected when a given digital code is applied and the actual output measured.

The overall accuracy is affected by quantizing, offset error, bandwidth, linearity, monotonicity, settling time and long term drift.

Acquisition time The time elapsed between the sample or track time command and the point at which the output tracks the input, in a S/H or T/H.

ADC Analog to digital converter.

A/D Same as ADC.

Aliasing Aliasing occurs when a signal is sampled at a rate less than twice per period of the highest frequency component of significant amplitude present. The signal is said to be *undersampled* and the result looks like a waveform of lower frequency.

Amplitude uncertainty (or **amplitude error**) This error is a function of the aperture uncertainty. It is expressed as $\Delta V = T_a dV/dt$, where ΔV is the amplitude uncertainty, T_a is the aperture uncertainty and dV/dt is the rate at which the signal is changing at the instant of sampling.

Aperture time delay The time elapsed between the hold command (or the end of the sample command) and the time at which the sampling switch is completely open.

Aperture uncertainty The variation in the aperture time delay.

Bandwidth The frequency at which the steady state sinewave response of a circuit or system has fallen by 3dB, i.e. is 70.7% of its value at lower frequencies. Limits the fastest risetime that a digital storage oscilloscope can accurately record.

DAC Digital to analog converter.

D/A Same as DAC.

Equivalent time bandwidth This is (usually) the highest bandwidth achieved by a digital storage oscilloscope. Uses random repetitive (random interleaved) sampling. Only achievable on repetitive signals.

Equivalent time sampling See *sampling*.

Glitch A pulse or spike, usually narrow, that occurs in a digital circuit or the output of a DAC. In digital circuits, commonly due to a timing violation. Occurrence may be limited to a rare combination of factors, making the glitch's appearance infrequent and unpredictable. Consequently, may have dire effects upon circuit operation and be difficult to troubleshoot.

Linearity A circuit is said to be linear when its output is strictly proportional to its input. In a DAC or ADC, implies that the analog voltage, plotted against the digital code, is a straight line. Error may be quoted as the maximum deviation of points from a best straight line fit, or from a straight line passing through the end points (e.g. zero and full scale).

Long term drift Variation in e.g. gain or offset of a system. Can affect all parameters of a system (except resolution and quantising error).

Glossary

Memory size Usually, the number of points or samples, per input channel, in a record or acquisition. Some digital storage oscilloscopes may be able to share channel memory, providing a longer record length when not all of the channels are in use. Others, with deep memories, may be able to segment them to store multiple acquisitions of a shorter length.

Monotonicity The output of a monotonic DAC or ADC never increases in response to a decreasing input stimulus, or vice versa.

Nyquist frequency If a continuous bandlimited signal contains no frequency components higher than f_n, then the original signal can in principle be completely recovered without distortion from a sampled version of the same provided that it was sampled at a rate greater than $2f_n$ samples per second. Here, f_n is referred to as the Nyquist frequency.

Offset error The output voltage of a DAC with the input code set to zero, or the dc input voltage required for an ADC to produce an output code of zero.

Perceptual aliasing An optical illusion caused by the eye's failure to perceive the correct time order of the displayed points when viewing a dot display. Occurs at frequencies much lower than a tenth of the sample frequency. Avoid by switching to linear interpolation (dot-joining) display mode.

Quantising error An ADC can only represent an analog input voltage as one of its digital output codes. Relative to an input voltage which converts exactly to a given output code, voltages in the range $-\frac{1}{2}$LSB to $+\frac{1}{2}$LSB will also be assigned to the same code. There is thus an inherent quantising uncertainty of $\pm\frac{1}{2}$LSB.

Realtime sampling See *sampling*.

Record length Usually defined as the number of samples that are acquired per record, per input channel.

Resolution The increment in input voltage corresponding to one LSB of ADC input range. For an *n*-bit binary converter, usually quoted as one part in 2^n, as a percentage of full scale, in parts per million, or as '*n* bits'.

Sampling

Equivalent time Used when the maximum sampling rate is inadequate to fill the acquisition memory in a single sweep (i.e. following a single trigger event). A number of samples are taken at various times during each sweep, until after sufficient sweeps the acquisition memory is full. Waveform of interest must be repetitive.

Randon repetitive sampling Same as *equivalent time sampling*.

Random interleaved sampling Same as *equivalent time sampling*.

Realtime sampling The sampling rate required does not exceed the maximum of the instrument. Consequently, all the samples required to form the record can be taken during a single sweep (i.e. following a single trigger event).

Recurrent sampling Operation with new acquisitions following one after another.

Single shot sampling Real time sampling with a single acquisition captured following the occurrence of a trigger event, or acquisition terminated on (or a defined time after) the occurrence of a trigger event.

Roll mode sampling Real time sampling with each new sample displayed at the right hand side of the screen as it is taken, all other samples being shunted one point to the left. Acquisition is continuous, like a chart recorder, or may be terminated on (or a defined time after) the occurrence of a trigger event.

Sample and hold (S/H) Acquisition method in which a sampling gate passes a very short sample of the instantaneous input signal to its output. An S/H is normally left in the HOLD condition.

Sample rate The frequency with which an analog signal is sampled, expressed in samples/second or Hertz. Limited by the maximum conversion time of the ADC (including the settling time of the S/H, if used).

Glossary

Sampling theorem If a continuous bandlimited signal contains no frequency components higher than f_n, then the original signal can in principle be completely recovered without distortion from a sampled version of the same provided that it was sampled at a rate greater than $2f_n$ samples per second. Here, f_n is referred to as the *Nyquist frequency*.

Settling time The time taken for the output of a DAC to settle, for a full scale code change, usually to within the analog equivalent of $\pm {}^1/_2$ LSB.

Track and hold (T/H) Acquisition method in which the output of a sampling gate tracks the instantaneous input signal, until switched to the hold condition. A T/H is normally left in the track condition.

Trigger mode The selected method of triggering:

> **Coupling** May be dc, ac, ac(LF reject), ac(HF reject)
>
> **Source** May be internal (channel 1, channel 2 etc), external, or line
>
> **Mode** slope (level, polarity); recurrent or single shot (arm, hold); direct or delayed (by specified time or (number of) event(s)) etc

Appendix

Here is a list of manufacturers of digital storage oscilloscopes and their agents. The list includes manufacturers of dual mode oscilloscopes (realtime analog oscilloscopes with built-in digital storage and display facilities).

This information is believed to be correct at time of going to press. However, neither the author nor the publisher can accept any responsibility for errors or omissions. Note that where an agent is quoted, that agent is not necessarily the main or sole agent.

Tel = telephone, Fax = facsimile, Tx = telex.

Fluke

USA — Fluke Corporation, PO Box 9090, Everett, Wa. USA 98206 Tel (800) 443-5853; Fax (206) 356-5116

UK — Fluke (UK) Ltd., Colonial Way, Watford, Herts, WD2 4TT, UK Tel 01923 240511; Fax 01923 225067; Tx 934583

Gould

USA — Gould Instrument Systems, Inc., 8333 Rockside Road, Valley View, OH 44125-6100 USA Tel (216) 328 7000; Fax (216) 328 7400

Appendix

UK — Gould Instrument Systems Ltd., Roebuck Road, Hainault, Ilford, Essex , IG6 3UE. UK Tel 0181 500 1000; Fax (0181) 501 0116; Tx 263785

Other offices throughout the world.

Hewlett-Packard

USA — Hewlett-Packard Company, 1421 S. Manhattan Ave., Fullerton, CA 92631, USA Tel (714) 999-6700

UK — Hewlett-Packard Limited, Test and Measurement, Cain Road, Bracknell, Berks., RG12 1HN, UK Tel 01344 366666; Fax 01344 362852

Other offices throughout the world.

Hitachi

Japan — Hitachi Denshi, Ltd. 23-2 Kanda-Chuda-cho 1-chome, Chiyoda-ku, Tokyo, Japan. Tel (03)3255-8411; Fax (03)3257-1433/1434; Tx HECO J24178

USA — Hitachi Denshi America. Ltd. 150 Crossways Park Drive, Woodbury, New York 11797, USA Tel (516)921-7200; Fax 516-496-3718; Tx 510-221-1899

UK — Hitachi Denshi (UK) Ltd. 13/14 Garrick Industrial Centre, Irving Way, Hendon, London NW9 6AQ, UK Tel (81)202-4311; Fax 81-202-2451; Tx 27449

Other offices throughout the world.

Kikusui

Japan — Kikusui Electronics Corp. Nisso Dai — 15 Bldg. 8th Fl. 2-17-19, Sin-Yokohama, Kohoku-ku, Yokohama, 222, Japan. Tel (045) 475-1112; Fax (045) 475-1115; Tx J36475 KECJPN

USA — Represented by Marconi Instruments, 6 Pearl Court, Allandale, N.J. USA Tel 201934 9050

UK — Represented by Telonic Instruments Ltd., Toutley Industrial Estate, Toutley Road, Wokingham, Berks RG41 1QN. UK Tel 01734 786911; Fax 01734 792338

Digital storage oscilloscopes

Leader

Japan — Leader Electronics Corporation, 2-6-33 Tsunashima-Higashi, Kohoku-ku, Yokohama, 223, Japan. Tel 045-541-2123; Fax 045-544-1280

USA — Leader Instruments Corp. 380 Oser Avenue, Hauppage, New York 11788, USA Tel 1-516.231-6900; Fax 1-516-231-5295; Tx 510-227-9669 Leader Haup.

UK — Leader Instruments (Europe) Ltd., Raglan House, 8-24 Stoke Road, Slough, Berks SL2 5AG, UK Tel 01753 538022; Fax 01753 538528

LeCroy

USA — LeCroy Corp. 700 Chestnut Ridge Road, Chestnut Ridge, N.Y. 10977-6499 USA Tel (914) 425 2000; Fax (914) 425 8967; TWX. (710) 577 2832

UK — LeCroy Ltd., 27 Blacklands Way, Abingdon Business Park, Abingdon, Oxon. OX14 1DY, UK Tel (01235) 533114; Fax (01235) 528796

Metrix

France — Metrix S.A. Chemin de la Croix Rouge — BP 2030, 74010-Annecy Cedex, France. Tel 50 336262; Fax 50 336200

UK — Metrix S.A. Jays Close — Viables Estate, Basingstoke, Hants., RG22 4BW, UK Tel (01256) 311877; Fax (01256) 23659

Offices and agents throughout the world.

Nicolet

USA — Nicolet Instrument Technologies, 5225-4 Verona Road, Madison, Wisconsin 53711-4495, USA Tel 608 276-5600; Fax 608 273-5061

UK — Nicolet Technologies Ltd. 26 Rockingham Drive, Linford Wood, Milton Keynes, MK14 6PD, UK Tel (01926) 494111; Fax (01926) 494452

Offices in France and The Netherlands

Appendix

Tektronix

USA — PO Box 1700 Beaverton, Oregon 97075 USA Tel (800) 835-9433, TWX: (910) 467-8708, TLX: 151754

UK — Fourth Avenue, Globe Park, Marlow, Bucks. SL7 1YD. Tel (01628) 403300, Fax: 01628 403403

Other offices throughout the world.

Yokogawa

Japan — Yokogawa Electric Corporation, Test and Measurement Sales Div., Shinjuku-Nomura Bdlg. 1-26-2 Nishi-Shinjuku, Shinjuku-ku, Tokyo, Japan. Tel 81-3-3349-1013; Fax 81-3-3349-1017

USA — Yokogawa Corporation of America, 2 Dart Road, Shenandoah Industrial Park, Newnan, GA. 30265, USA Tel 404-253-7000; Fax 404-251-2088

UK — Represented by Martron Instruments Ltd., Wellington Road, High Wycombe, Bucks HP12 3PR. Tel 01494 459200; Fax 01494 535002

European Office in The Netherlands

Index

A

ADC (analog to digital converter) 7, 9, 11, 19, 21, 22, 30, 34–40, 47, 54, 58, 63, 98, 133
   successive approximation register (SAR) 34
   flash 34, 35
A/D see ADC
alias, -ing 4, 30, 68, 84, 120–123, 133
   anti ~ algorithm 122
   perceptual ~ 30, 66, 75, 78, 79
amplifier
   buffer ~ 35
   Y output ~ 69, 71, 72, 86
   X output ~ 71, 72
amp-second product 113
arming 10, 17, 42
attenuator 27
   Y input ~ 11, 21, 24, 30, 33, 86
averaging 49–66, 101, 133
   exponential ~ 53
   stable ~ 53, 54
auto
   ~set-up, -ranging 28, 30, 123

B

babysitting 58
bandwidth 27, 50, 63, 77, 80, 86, 101, 135
   analog ~ 25
   single shot or real time ~ 21
BER see bit error rate
bit 33, 36
   least significant ~ (LSB) 37
   most significant ~ (MSB) 34, 37
   next ~ ~ ~ (NMSB) 34
   ~ error rate 61, 62
blanking
   inter-dot ~ 93
bucking current 114
blow-by 98, 99
byte 33

C

calibration, calibrators 118
capacitor, capacitance
   hold ~ 35
   stray ~ 27
carrier 57
Centronics interface 103
charge coupled devices (CCDs) 38
chart recorder 10
clock(s)
   recovered ~ 61
   staggered ~ 39
common mode rejection ratio (CMRR) 137
comparator 34
   straobe ~ 90, 94

Index

conversion, converter
   start ~ command 36
   sub-ranging ~ 36
coupling
   dc ~, ac ~ 28
   fibre optic ~ 136, 137
   opto ~ 136
   transformer ~ 136
CRT (cathode ray tube) 7, 9, 15, 25, 86
   magnetic deflection ~ 8, 59, 67–83
   electrostatic deflection ~ 59, 68
cursors 9, 72

D

DAC (digital to analog converter) 7, 14, 34, 71, 98
   multiplying ~ 75
dead time 20, 126
decimation 133, 134
deflection
   ~ plates
      X ~ ~ 69
      Y ~ ~ 69
   Y ~ 75
delay
   group ~ 123
   ~ line 89, 91, 101, 102
      sampled analog clocked ~ ~ 38
detectors
   analog peak ~ 58
differential
   ~ amplifier 136
   ~ probes 3
digital
   ~ signal processing 61
digitising 37
display
   bit-mapped ~ 60, 62
   electroluminescent ~ 67
   vector scanned ~ 69, 71
dot-joining see interpolator, pulse
drift 30
droop 37
DSP see digital signal processing
dual-mode oscilloscopes 1, 7, 41, 59, 68, 123

E

equivalent time (repetitive mode) 11, 21–25, 63, 84, 127
expansion
   X ~, Y ~ 135
   post-storage ~ 30, 34
eye diagrams 60–65

F

factor
   deflection ~ 30
FET 28
filter
   anti-alias ~ 59, 124
   ~ corner, breakpoint or cut-off frequency 45
   Gaussian 123
   high-, low-pass 32, 45
   linear phase 134
flyback 10, 20

G

glitch 3, 20, 22, 45–47, 58, 101, 125, 133
   ~ stretching 48
GO/NO GO testing, facility 44
graticule 14, 21, 22, 30, 71
grounding
   indirect ~ 115

H

Hall effect device 113
histogram 63
holdoff 20, 90, 125
HPIB interface 103

I

impedance, low, high 26
   characteristic ~ 27
   input ~ 28
interpolator
   ~ pulse 73, 78, 80, 122
   sine ~ 21, 32, 63–66, 79–82
isolators (voltage, ground) 136–138
inverters 118

149

J

jitter 80, 102, 135

K

kickout 92, 94, 96, 98

L

LCD (liquid crystal display) 7, 67–70
linearity 9
logic
    ~ analys-er, -is 5, 45
    ~ decode- 35
long-tailed pair 28

M

mains isolation 114–118
memory 14, 16, 92
    acquisition ~ 18, 20, 22, 23, 33
    display ~ 18, 20, 49
    ~ gate 95
mode
    dot-joining ~ 56
    envelope ~ 44, 47, 56 -58, 122, 133
    MAX/MIN ~, see ~, envelope
multiplex-er, -ing 39
    de~ 39

N

Nyquist rate, criterion 31, 32, 56, 63, 122, 123, 131, 133, 135
nibble 33
noise 54–56, 61, 100, 101, 133
    high frequency ~ 50
    random ~ 54
    sampling ~ 86
    white ~ 133
notation
    binary ~ 33
    HEX ~ 33

O

offset 28
over-(under)shoot 97

P

persistence
    infinite ~ 44, 60
phase
    ~ lag, ~advance 45
pixel 60, 62
plotter 12
post trigger 17
pretrigger 11, 17, 18, 43
probes 3, 12, 85, 89, 103–114
    active ~ 109–111
    FET input ~ ~ 102, 110, 111
    current ~ 110–114
    differential ~ 107, 109
    passive ~ 105–109
processing
    per channel ~ 49
    post ~ 49
    pre ~ 21, 49, 64
        ~ digitising ~ 58
    waveform ~ 49–66
pulse
    ~ amplitude modulation (PAM) 38
    runt ~ 48, 127
    ~ width 47
    strobe ~ 92

Q

quantis-ation, -ing, -ed 31, 38

R

RAM (random access memory) 7, 16, 20, 39
raster 8, 25
    ~ scan 59, 67–71
rate
    repetition ~ 24
record (length) 131, 132
recurrent mode 10, 18
refresh mode 10, 18–21
repetitive (equivalent time) mode 12, 63
resolution 34, 54, 133, 135
retrace 10, 20, 125
ringing 81
risetime 77, 81–83, 101

Index

roll mode 2, 9, 13–17,
RS232 103
R/W (read/write) 16, 20

S

sample-and-hold (S/H) 25, 35, 36
sample, -ing
~ clock 22, 23, 24
~ feedback loop (SFBL) 98
equivalent time ~ 8, 23
~ gate 24, 89–95
~ head 89, 90
~ oscilloscope 84 -102
~ pulse 90–94
random ~ 101, 102
random repetitive or interleaved ~ 21, 128, 130, 135
~ ~ multiple point ~ 24, 25
~ rate 10, 11, 18, 30
sequential ~ 24, 87, 101, 128, 130
scan
over ~ 30
screen (display) 16, 18, 19, 22
SCSI interface 104
sensitivity
variable ~ 27
shift (control), X, Y 30, 43
signal
amplitude modulated ~ 57
~ condition-er, -ing 116
~ isolation 115, 118, 136–138
~ to noise ratio 50, 54, 55
~ processing 56
single shot 10, 11, 17, 18 , 25, 38, 50
smoothing 49–66, 133
spectrum 60, 62
stroboscope 86, 87

T

threshold 11
time
acquisition ~ 36
aperture ~ 24, 36
display ~ 32
~ domain 38
settling ~ 36
timebase 10, 18, 32, 123
dual ~ (A and B) 70
trace 14
track-and-hold (T/H) 36
transducer 14
transient 2
trigger 10, 16, 30, 87–89, 90, 102
auto ~ 18, 42
~ countdown 89
~ coupling 43
~ ~ HF reject 52
~ delay
~ ~ by time or event 48
edge ~ 125
hysteresis ~ 45, 46
~ level 18, 43, 45, 47, 90
normal ~ 18, 42
single shot ~ 42
~ slope 90
TV ~, line, frame 43
window ~ 45, 46
-ing 42–47
internal, external- 42
line ~ 42
pattern ~ 48

V

vector 73
vector scanned display, see display
volt-second product 112, 119

W

waveform
sawtooth ~ 75
window 15
display ~ 30

Y

Y
~ input 9, 16, 30
~ ~ (pre)amplifier 11, 26, 27, 33, 35, 63
balanced ~ ~ ~ 30

151